Contents

PLANTS OF
EAST SABAH

R. P. J. de Kok and T. M. A. Utteridge

Kew Publishing
Royal Botanic Gardens, Kew

PLANTS PEOPLE
POSSIBILITIES

DARWIN
INITIATIVE

First published in 2010 by
Royal Botanic Gardens, Kew
Richmond, Surrey, TW9 3AB, UK
www.kew.org

ISBN 978 1 84246 378 9

British Library Cataloguing in Publication Data
A catalogue record for this book is available from the British Library.

Production editor: Sharon Whitehead
Typesetting and page layout: Margaret Newman
Publishing, Design & Photography
Royal Botanic Gardens, Kew

Cover design: Margaret Newman
Front cover photograph: Maliau Basin, Sabah (© Andrew McRobb/RBG Kew)

Printed and bound in the United Kingdom by Henry Ling Ltd

For information or to purchase all Kew titles please visit
www.kewbooks.com or email publishing@kew.org

Kew's mission is to inspire and deliver science-based plant conservation worldwide, enhancing
the quality of life.

All proceeds go to support Kew's work in saving the world's plants for life.

Mixed Sources
Product group from well-managed
forests and other controlled sources
www.fsc.org   Cert no. SA-COC-001860
© 1996 Forest Stewardship Council

FSC

The paper used in this book contains wood from
well-managed forests, certified in accordance with the
strict environmental, social and economic standards
of the Forest Stewardship Council (FSC).

Introduction

The production of this field guide and checklist was part of a UK government, Darwin Initiative sponsored project 'Assessing and conserving plant diversity in commercially managed tropical rainforests'. This project was a joint initiative between the Royal Botanic Gardens, Kew, Yayasan Sabah and the Royal Society's South East Asia Rainforest Research Programme, and was assisted by the Forest Research Centre Herbarium of the Sabah Forestry Department.

In this guide, the 84 most commonly encountered families in the lowland rainforest of Danum Valley, Maliau Basin, Imbak Canyon and the areas in between (see Map 1) are described. The selection of plant families for inclusion in this field guide was based mainly on the number of specimens from the area in the database held at the herbarium of the Sabah Forestry Department, Sandakan in early 2009.

The field characters described in this book should make it possible to identify the vast majority of specimens to family level. A more detailed description of each family is also provided as a backup. Families with which each described family are most usually confused are listed alongside the key characters that distinguish them. In almost all cases, genera within the described family are listed (sometimes all genera, more usually the more common ones in East Sabah) together with some of their distinguishing characteristics.

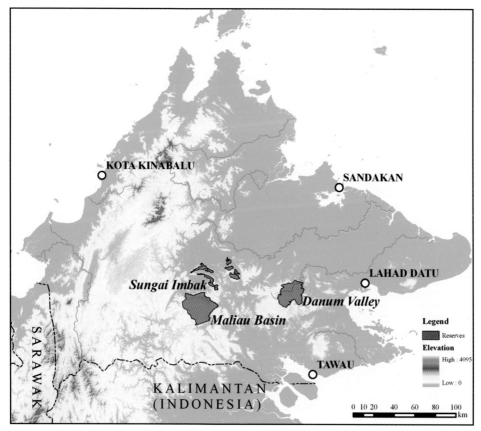

Map 1. Sabah

Opposite: © T. Utteridge

Resources for use with this book

There are many publications to help you get to know plants from Sabah. The easiest to use is Lee's (2003) *Preferred Check-list of Sabah Tees*, which has a small list of key characters at the back of the book. The series of books on Malesian Seed Plants by van Balgooy (1997, 1998, 2001) are very useful and easy to use for indentifying plants to family and often to genus level. The *Interactive Key to Malesian Seed Plants*, which is freely available on the web (www.kew.org/herbarium/keys/fm) is largely based on these books. This key is a good way of identifying specimens to family level and has many pictures.

Once you know which family your specimen belongs to, there are many different publications to help you further. The *Trees of Sabah* (Cockburn 1976, 1980), the *Tree Flora of Sabah and Sarawak* series (Soepadmo & Wong (1995), Soepadmo *et al.*, 1996 and ongoing) and the *Tree Flora of Penisular Malaysia* (Whitmore, 1972; Ng, 1978–1989) are very useful. The *Flora Malesiana* series (van Steenis, 1950–2007) and the *Flora of Java* (Backer & Bakhuizen van den Brink, 1963–1968) are more difficult to use, but are the only source of ready available information for many plant groups. Many different groups have been described in different publications, either in books or journals. Some of these publications will be mentioned in van Balgooy (1997, 1998, 2001), in the *Interactive Key to Malesian Seed Plants*, or in the various online catalogues such as that of the library of The Royal Botanic Gardens, Kew (www.kew.org/library), the Kew Record of Taxonomic Literature (http://kbd.kew.org/kbd/searchpage.do) or the World Checklists of Monocotyledons (www.kew.org/wcsp/monocots).

Literature

Backer, C. A. & Bakhuizen van den Brink, R. C. (1963–1968). Flora of Java, Vol. I–III. Noordhoff, Groningen.

van Balgooy, M. M. J. (1997). Malesian Seed Plants, Vol. 1 — Spot Characters. Rijksherbarium/Hortus Botanicus, Leiden.

van Balgooy, M. M. J. (1998). Malesian Seed Plants, Vol. 2 — Portraits of Tree Families. Rijksherbarium/Hortus Botanicus, Leiden.

van Balgooy, M. M. J. (2001). Malesian Seed Plants, Vol. 3 — Portraits of Non-tree Families. Nationaal Herbarium Nederland–Universiteit Leiden Branch, Leiden.

Cockburn, P. F. (1976). Trees of Sabah, Vol. I. Forest Department Sabah, Sandakan.

Cockburn, P. F. (1980). Trees of Sabah, Vol. II. Forest Department Sabah, Sandakan.

Lee, Y. S. (2003). Preferred Check-list of Sabah Tees, 3rd Edn. Natural History Publications (Borneo) Kota Kinabalu and Sabah Forestry Department, Sandakan.

Ng, F. S. P. (1978–1989). Tree Flora of Malaya, Vol. III–IV. Longman, Malaysia.

Soepadmo, E. & Wong, K. M. (eds) (1995). Tree Flora of Sabah and Sarawak, Vol. 1. Sabah. Forestry Department, Forest Research Institute Malaysia & Sarawak Forestry Department, Sandakan.

Soepadmo, E., Wong, K. M. & Saw, L. G. (eds.) (1996). Tree Flora of Sabah and Sarawak, Vol. 2. Sabah Forestry Department, Forest Research Institute Malaysia & Sarawak Forestry Department, Sandakan.

van Steenis, C. G. G. J. (ed.) (1950–2007). Flora Malesiana Vol. 1–18. National Herbarium, The Nederlands.

Whitmore, T. C. (1972). Tree Flora of Malaya, Vol. I & II. Longman, Malaysia.

Online resources

Royal Botanic Gardens, Kew. Interactive Key to Malesian Seed Plants. Available online: http://www.kew.org/herbarium/keys/fm/

Royal Botanic Gardens, Kew. Kew Library Catalogue. Available online: http://www.kew.org/library/

Royal Botanic Gardens, Kew. Kew Record of Taxonomic Literature. Available online: http://kbd.kew.org/kbd/searchpage.do

Royal Botanic Gardens, Kew. World Checklists of Monocotyledons. Available online: http://www.kew.org/wcsp/monocots

Acanthaceae

Field characters:

Herbs or lianas, stems with swollen nodes when fresh (shrunken ones when dry); leaves with cystoliths usually present; corolla tubular; usually 4 stamens; fruit a capsule, usually with hooks inside; seeds few.

Description:

Habit herbs, sometimes shrubs or rarely lianas.
Sap absent.
Stipules absent.
Stem with swollen nodes when fresh or shrunken nodes when dried.
Leaves opposite, decussate, simple, margins entire to serrate, usually with cystoliths present.
Inflorescence determinate; flowers often associated with showy bracts.

Flowers corolla funnel-shaped, 2-lipped or 1-lipped.
Ovary superior, ovules usually 2–10.
Fruit 2-valved, often a stipulate capsule, usually with jaculators (hooks) inside; seeds few.

Confused with:

Gesneriaceae
Leaves without cystoliths, often hairy; fruit a capsule or indehiscent berry without hooks; seeds many, tiny.

Lamiaceae
Often with angular stems; cystoliths absent; fruit usually a drupe; 1–4 seeds.

Scrophulariaceae
Cystoliths absent; fruit usually with persistent style; seeds many.

© G. Bramley

Strobilanthes leaves with cystoliths

© R. de Kok

Swollen nodes on stem (nodes are shrunken when dry)

© G. Bramley

Ruellia fruit with hooks

© T. Utteridge

Ruellia flower

Major genera:	
Acanthus	Herb with spines, cystoliths absent; corolla with upper lip absent; fruit with hooks inside.
Barleria	Herb with spines, cystoliths present; calyx with 2 large and 2 small lobes; fruit with hooks inside.
Justicia	Herb without spines, cystoliths present; filaments 2, with the 2 thecae attached at different levels; fruit with hooks inside.
Ruellia	Herb without spines, cystoliths present; corolla lobes sub-actinomorphic; fruit with hooks inside; seeds more than 4.
Staurogyne	Small herb without spines, cystoliths absent; corolla lobes sub-actinomorphic; fruit without hooks inside.
Strobilanthes	Herb without spines, with cystoliths present; inflorescences large; fruit with hooks inside.
Thunbergia	Liana, cystoliths absent; flowers surrounded by two big bracts; fruit duckbill-shaped, without hooks inside; seeds globose.

Actinidiaceae

Field characters:

Trees, shrubs and lianas; leaves simple; stipules absent; flowers regular, 5-merous, sepals free, petals fused at base; stamens numerous and fused with the base of corolla; ovary superior, styles free; fruit a berry.

Confused with:

Dilleniaceae
Stipules present; petals free, carpels free; seed arillate.

Theaceae
Leaves not scaly; flowers solitary with a pair of bracteoles; fruit a berry or a dry capsule.

Description:

Habit trees, shrubs or lianas; sometimes with spines.

Sap absent.

Stipules absent.

Leaves usually spirally arranged, simple, penninerved, margins often dentate, often covered with scales or stellate hairs.

Flowers regular, 5-merous, usually bisexual; sepals free, petals fused at base; stamens numerous and fused with base of corolla.

Ovary superior, styles free.

Fruit a berry.

© R. de Kok

Leaves with scales of *Saurauia agamae*

© RBG Kew

Saurauia sp.

© J. Gregson

Flower of *Saurauia* sp.

Major genera:	
Actinidia	Lianas or climbing shrubs; leaves without scales, but with stellate or simple hairs; styles 5 to numerous.
Saurauia	Trees or shrubs; leaves often covered with scales or broad hairs; single style, often branched.

Alangiaceae

Field characters:

Trees; leaves simple, 3-veined to palmately veined; stipules absent; inflorescence cymose; flowers 4–10-merous; petals free, valvate; disk intra-staminal; ovary inferior.

Description:

Habit trees, rarely shrubs or lianas.

Sap absent.

Stipules absent.

Leaves alternate or spirally arranged, simple, margins usually entire, usually unequal leaf bases, venation more or less palmate, sometimes 3-veined from base, domatia present, sometimes with stellate hairs, petiole sometimes with swollen apices.

Flowers 4–10-merous, bisexual; petals free, valvate, usually white; stamens usually twice the number of petals, intra-staminal disk.

Ovary inferior, 1–2 locular, 1 style present.

Fruit a drupe with persisting calyx tube at apex, sometimes ridged or latterly compressed.

Confused with:

Cornaceae

Leaves often decussate; stamens as many as petals.

Alangium javanicum

Flower of *Alangium javanicum*

Genus:

Alangium   As for the family.

Anacardiaceae

Field characters:

Trees, shrubs or lianas; sap white, drying black after exposure, often causing irritation; leaves simple or compound, often obovate; stipules absent; fruit usually laterally compressed.

Description:

Habit trees, shrubs or lianas; usually with an acidic smell coming from crushed leaves.

Sap white sap that dries black or brown from cut surfaces, may cause painful irritation of the skin.

Stipules absent.

Leaves alternate (spirally arranged or whorled), very rarely opposite, simple, pinnate or trifoliolate, often crowded near apex; margins entire; petiole often swollen at the base.

Flowers small, usually 4–5(–6)-merous; disk present; anthers 10–15, borne outside or rarely on the nectary disk.

Ovary superior or occasionally (semi-)inferior, many-locular or rarely 1-locular, always with 1 ovule per cell; style often eccentric.

Fruit usually a drupe, sometimes winged or on an enlarged fleshy pedicel and receptacle; seeds 1–5(–12).

Confused with:

Burseraceae
Usually no black or white sap present; flowers often 3-merous; 2 ovules per cell.

Fruit of *Semecarpus* sp.

© J. Dransfield

Major genera:

Buchanania	Trees; leaves spirally arranged, simple; flowers 4–6-merous; ovary superior, style 1, short; fruit a drupe.
Campnosperma	Trees with *Terminalia*-type branching; leaves spirally arranged, simple, with peltate scales; flowers (3–)4(– 5)-merous; ovary superior, style and stigma 1; fruit a drupe.
Dracontomelon	Tree; leaves spirally arranged, pinnate with <10 leaflet pairs; flowers 5-merous; ovary superior, styles 5; fruit a drupe.
Gluta	Trees or shrubs; leaves spirally arranged or in whorls, simple; flowers and stamens 5(–8)-merous; ovary superior, style and stigma 1; fruit a drupe, sometimes with wings.
Koordersiodendron	Trees; leaves spirally arranged, compound with >10 leaflet pairs; flowers 5-merous; ovary superior, styles 5; fruit a drupe.
Mangifera	Trees; leaves spirally arranged, simple; flowers 4–5-merous; ovary superior, style and stigma 1, usually lateral; fruit a glabrous drupe.
Melanochyla	Trees; leaves spirally arranged, simple; flowers (4–)5-merous; ovary superior or inferior, style 1, stigmas 3; fruit a hairy drupe.
Pegia	Climber; leaves alternate, compound with <4 leaflet pairs; flowers (4–)5-merous; ovary superior to semi-inferior, styles 4–5; stamens twice as many as petals; fruit a drupe.
Semecarpus	Trees or shrubs; leaves spirally arranged or whorled, simple; flowers (4–)5-merous; ovary superior or inferior, styles 3; fruit a drupe with enlarged fleshy pedicel and receptacle.

Annonaceae

Field characters:

Trees, shrubs or climbers; twigs dark striate, cross-section of wood with rays; leaves alternate, entire; stipules absent; flowers 3-merous; stamens numerous; carpels numerous.

Description:

Habit trees, shrubs or climbers; twigs often dark with a striate surface, cut cross-section shows medullary rays in the wood.

Sap usually absent.

Stipules absent.

Leaves alternate, usually distichous, rarely spirally arranged, simple; margins entire; hairs where present, simple or stellate or with lepidote scales.

Inflorescences or **flowers** axillary, terminal or opposite the leaves; ramiflorous or cauliflorous.

Flowers 3-merous; sepals and petals present; valvate or imbricate in bud; stamens numerous; anthers linear.

Ovary superior, 1 to numerous, free.

Fruit consisting of 1 to several fleshy indehiscent monocarps or syncarpous; seeds many in 1-locular carpels, or 1-seeded.

Confused with:

Ebenaceae
Cross-section of wood without rays; flowers 4–5-merous; carpels fused.

Magnoliaceae
Cross-section of wood without rays; stipules present; flowers with many petals.

Monimiaceae
Leaves opposite, margins dentate.

Myristicaceae
Red sap present; cross-section of wood without rays; flowers tiny with a single carpel.

Uvaria javana

Fruit of *Uvaria* sp.

Polyathia sumatrana

Cross-section of twig of *Artabotrys* sp.   © R. de Kok

Hooks of *Artabotrys roseus*   © J. Gregson

Major genera:

Anaxagorea	Small trees or shrubs; inflorescence terminal or extra-axillary, sometimes opposite leaves; flowers with valvate petals; fruit with dehiscent clavate follicles.
Artabotrys	Lianas with hooks; inflorescence extra-axillary, sometimes cauliflorous or opposite leaves; flowers with valvate petals and a projecting rim where the blade joins the claw.
Cyathocalyx	Monopodial trees; inflorescence extra-axillary, sometimes opposite leaves; leaves sometimes with stellate hairs; flowers with valvate petals.
Cyathostemma	Climbers; inflorescence extra-axillary or cauliflorous; flowers with imbricate petals.
Fissistigma	Climbers; inflorescence terminal, extra-axillary or opposite leaves; flowers with valvate petals.
Goniothalamus	Small trees and shrubs; inflorescence axillary, sometimes terminal, or cauliflorous; flowers with valvate petals.
Monocarpia	Tall trees; leaves with intermarginal veins; inflorescence extra-axillary, sometimes opposite leaves; flowers with valvate petals; fruit with tubercles.
Orophea	Shrub or small trees; inflorescence (extra-)axillary; flowers with valvate petals; fruit sometimes moniliform.
Phaeanthus	Shrub or small trees; inflorescence extra-axillary; leaves turning black when dried; flowers with valvate petals.
Polyalthia	Trees and shrubs; inflorescence terminal, (extra-)axillary or cauliflorous; flowers with valvate petals; stamens with appendages.
Pseuduvaria	Trees and shrubs; inflorescence axillary; flowers with valvate petals; fruit with tubercles.
Uvaria	Lianas and scramblers; sometimes with red sap; usually with stellate hairs; inflorescence terminal or opposite leaves, sometimes cauliflorous; flowers with imbricate petals.
Xylopia	Trees; inflorescence axillary; flowers with valvate petals; fruit monocarps moniliform and dehiscent.

Apocynaceae

Field characters:

Plants often rather fleshy; milky sap present; leaves opposite, sometimes with ladder-like secondary veins, stipules absent; corolla lobes distinctly overlapping; ovary superior; fruit paired; seeds with a tuft of hairs at the apex.

Description:

Habit trees, shrubs or lianas.

Sap white, sometimes clear.

Stipules absent.

Leaves opposite to whorled or rarely alternate; simple, margins entire.

Flowers bisexual, mostly 5-merous; corolla tubular, symmetrical; stamens 4–5, anthers often highly modified.

Ovary superior, carpels usually 2, usually fused only at base or at apex, apical portion of style expanded and forming a head.

Fruit generally paired or single via abortion; seeds normally with a tuft of hairs.

Confused with:

Asclepiadaceae
Herb, climbers or lianas; corolla with a corona; stamens united with style.

Guttiferae
Clear, black or yellowish sap present; leaves often with pellucid or black dots; stamens many.

Lamiaceae
Sap absent; leaf margins often serrate or dentate; fruit 1–4-lobed capsules.

Melastomataceae Sap absent; leaves with 2–8 basal nerves; anthers articulated or with appendages.

Myrtaceae
Sap absent; leaves with pellucid gland dots; flowers usually with many anthers.

Oleaceae
Sap absent; leaves often sub-opposite; petals 4, stamens 2.

Rhizophoraceae
Sap absent; interpetiolar stipules present; petals free; stamens number more than petals; ovary superior to inferior.

Rubiaceae
Sap absent; interpetiolar stipules present; ovary usually inferior.

© R. de Kok

Fruit of *Cerbera* sp.

© RBG Kew

Fruit of *Tabernaemontana* sp.

© RBG Kew

Flowers of *Alstonia* sp.

Major genera:

Alstonia	Shrubs or trees; leaves in whorls, parallel secondary venation; fruit a pair of follicles; seeds with a tuft of hairs.
Alyxia	Shrubs or climbers; leaves opposite or in whorls, parallel secondary venation; fruit a pair of drupes; seeds simple.
Anodendron	Climbers; leaves opposite; ovary glabrous; fruit a pair of follicles; seeds with a tuft of hairs.
Chilocarpus	Climber; leaves opposite, intermarginal veins, black glands on underside; fruit a berry or a capsule; seeds with a corky aril.
Ichnocarpus	Climber; leaves opposite; ovary pubescent; fruit a pair of follicles; seeds with a tuft of hairs.
Kibatalia	Trees or shrubs; leaves opposite; fruit a pair of follicles; seeds with hairs pointing towards the seeds.
Kopsia	Trees; leaves opposite; fruit a pair of drupes; seeds not with a tuft of hairs, usually with a spur.
Leuconotis	Trees or lianas; leaves opposite, intermarginal vein, black glands on underside; fruit a berry; seeds simple.
Parameria	Climber; leaves opposite, rarely whorled; fruit a pair of follicles; seeds winged.
Rauvolfia	Trees; leaves mostly in whorls; dried plants resinous; fruit a drupe; seeds laterally compressed.
Tabernaemontana	Trees; leaves opposite or in whorls; interpetiolar stipules present; fruit a pair of follicles; seeds covered in a fleshy aril.
Urceola	Climber; leaves opposite; fruit a pair of follicles; seeds hirsute, flattened, with a tuft of hairs.
Willughbeia	Climber with tendrils; leaves opposite; fruit fleshy berry; seeds compressed.

Araceae

Field characters:

Herbs, climbing, epiphytic or terrestrial; leaves usually petiolate, with a compound midrib usually with pinnate branches; inflorescence scapose, with a spathe and spadix, other bracts absent; rarely floating aquatic herbs.

Description:

Habit herbs, climbing, terrestrial or epiphytic or rarely aquatic.

Sap absent, rarely clear.

Stipules absent.

Hairs sometimes with trichomes, scales, prickles or warty outgrowths.

Leaves alternate or spirally arranged, usually divided into a blade, petiole and petiole sheath; deeply lobed or compound; often leathery; margins entire; midrib compound, primary venation usually pinnately branched but sometimes palmate, curved or parallel; secondary venation reticulate or parallel-pinnate.

Inflorescences scapose, with a spathe (a solitary specialised bract) and a spadix (a dense spike of small flowers), other bracts absent.

Flowers very small, bi- or unisexual, when unisexual the female flower below the male on the spadix.

Ovary superior.

Fruit a berry.

Confused with:

Lemnaceae
Small floating herbs.

Marantaceae
Leaves alternate, with an open sheath; petiole swollen apically.

Orchidaceae
Stamens and style forming a column; roots covered in a velvet sheath.

Taccaceae
Inflorescence umbellate; flowers large.

Zingiberaceae
Aromatic leaves alternate, with an open sheath; ligule present; flowers large.

© R. de Kok

Alocasia sp.

© R. de Kok

Inflorescence of *Amorphophallus* sp.

© J. Gregson

Anadendrum microstachyum

Major genera:

Aglaonema	Climbing; leaves with striate venation; petiole short.
Alocasia	Terrestrial, tuberous; leaves sometimes peltate, collecting vein present.
Amorphophallus	Terrestrial, tuberous; often leafless when flowering; leaves single, deeply lobed with reticulate venation.
Colocasia	Terrestrial, tuberous; leaves sometimes peltate; collecting vein present; cut surfaces turning orange-brown.
Homalomena	Terrestrial, rarely climbing; aromatic when crushed; leaves sometimes peltate, with striate venation.
Pothos	Climbing; leaves with double intramarginal vein, often with winged petiole.
Rhaphidophora	Climbing; leaves often with holes, cystoliths present.
Schismatoglottis	Climbing; leaves with striate venation, apical ligule present.
Scindapsus	Climbing; leaves often with holes, joint at apex.
Typhonium	Terrestrial; leaves with reticulate venation; stamens with appendages.

Araliaceae

Field characters:

Leaves spirally arranged, often palmately or pinnately compound; ultimate inflorescence branch-umbellate; ovary inferior.

Description:

Habit trees, shrubs or climbers; sometimes spiny.

Sap absent, sometimes with a little clear sap.

Stipules often present, sometimes intrapetiolar or petioles crested at the base.

Leaves spirally arranged, rarely opposite; usually palmately or pinnately compound, sometimes bi- or tripinnate, rarely unifoliolate; margins entire to toothed.

Inflorescences terminal, rarely axillary; usually umbels or heads arranged in branched complexes.

Flowers usually bisexual, 5-merous; calyx absent or present, teeth very small; petals free or connate at the base only, valvate or imbricate.

Ovary inferior, 2–5 locules, rarely many.

Fruit drupaceous with 1 seed in each pyrene, or a berry, rarely a schizocarp.

Confused with:

Apiaceae (Umbelliferae) Never woody in Borneo; leaves usually simple, dissected.

Cornaceae Leaves opposite, always simple; stipules absent; flowers not in umbels.

Rosaceae–Prunus Leaves simple, glands on blade and/or petiole; ovary superior.

© R. de Kok

Intrapetiolar stipules of *Schefflera* sp.

© T. Utteridge

Flowers of *Schefflera* sp.

© R. de Kok

Leaves and inflorescence of *Osmoxylon* sp.

Major genera:	
Aralia	Shrubs or small trees, often armed; leaves pinnate to tripinnate; stipules absent but petioles with a sheathing base; petals imbricate.
Arthrophyllum	Trees or shrubs, with brown hairs; leaves crowded at the end of branches; inflorescences on leafy branches; petals valvate.
Gastonia	Trees, unarmed; leaves imparipinnate; stipules absent; pedicel not articulate below flower; petals valvate.
Osmoxylon	Shrubs or small trees, unarmed; leaves palmately lobed or simple, rarely digitately compound; stipules present; petiole with crest at base; inflorescence trifid; petals valvate.
Polyscias	Shrubs or small trees, unarmed; leaves imparipinnate, sometimes 2–3-pinnate or unifoliolate; stipules absent but with sheathing base; pedicels articulate below flowers; petals valvate.
Trevesia	Shrubs or lianas, sometimes unarmed; leaves palmately compound, petiolules sometimes joint together by a web of tissue; stipules present; petals valvate.
Schefflera	Trees, shrubs or lianas, unarmed; leaves unifoliolate (then articulation between petiole and blade) or palmately compound; stipules present; petals valvate.

Araucariaceae

Field characters:

Trees; leaves opposite or spirally arranged, simple, margins entire; flowers in unisexual cones.

Description:

Habit trees, rarely shrubs.

Sap resinous, sometimes white.

Stipules absent.

Leaves opposite to spirally arranged or whorled; simple, scale or needle-like to broad-leaved, margins entire; lamina with many faint closed longitudinal veins.

Flowers in unisexual cones, male and female cones on separate branches; male cones usually axillary on branches and cylindrical; female usually terminal on branches and ovoid to round.

Seeds dry, sometimes winged.

Confused with:

Casuarinaceae
Trees; twigs jointed; flowers with a proper perianth.

Gnetum
Trees to lianas; twigs with swollen nodes.

Podocarpaceae
Trees, irregularly branched; leaves sometimes scale-like; cones reduced (often fleshy) with few wingless seeds.

© R. de Kok

Cone of *Agathis borneensis*

Genus:	
Agathis	Leaves opposite, broad, with parallel venation; seeds winged.

Asclepiadaceae

Field characters:

Climbers or lianas; usually with milky sap; leaves opposite, simple, margins entire; ovary superior; fruit two follicles, seeds many.

Description:

Habit Climbers, epiphytes herbs or lianas.

Sap usually present, white.

Stipules absent.

Leaves opposite, simple, margins entire, usually with a tuft of trichomes at the base.

Flowers corolla with appendages forming a corona; stamens 5, united with the style.

Ovary superior, two free carpels united by the style.

Fruit two follicles; seeds many, often hairy or flattened.

Confused with:

Acanthaceae

Herbs; sap absent; leaves usually with cystoliths; flowers with 2 or 4 free stamens; fruit a capsule often with internal hooks.

Apocynaceae

Trees, shrubs or lianas; corolla usually large, corona absent; stamens not united with style.

Lamiaceae

Herbs to trees, rarely lianas, sap absent; flowers with 2 or 4 free stamens; fruit drupe-like, 1–4 locular.

Rubiaceae

Herbs to trees, rarely lianas, sap absent; stipules present, interpetiolar; ovary inferior.

Hoya flowers

© J. Dransfield

Asclepias currassavica flowers

© J. Gregson

Asclepiadaceae cont.

© J. Gregson

Asclepias currassavica fruit and seeds

Major genera:	
Asclepias	Herbs, often woody at base; leaves herbaceous, without a tuft of trichomes at base; filaments connate into a tube.
Dischidia	Epiphytic herbs, often climbing; leaves fleshy, sometimes formed into pitchers; filaments shortly connate at base.
Hoya	Herbs or shrubs, often epiphytic; leaves fleshy, sometimes with a tuft of trichomes at the base; filaments shortly connate at base.
Tylophora	Herbs or shrubs; leaves herbaceous, usually with a tuft of trichomes at the base; filaments shortly connate at base.

Asteraceae

Field characters:

Usually herbs, sometimes shrubs or trees; stipules absent; flowers in a head surrounded by bracts; calyx often replaced by a pappus of scales or bristles; petals fused; ovary inferior; style bifid.

Flowers of *Olearia* sp.

Description:

Habit herbs or shrubs, rarely trees or lianas.

Sap usually absent, white when present.

Stipules absent.

Leaves alternate or opposite; usually simple, rarely compound; margins entire to serrate.

Flowers gathered in a head-like capitulum consisting of flowers (florets) surrounded by a series of protective bracts; florets consist of ray- (absent in 50% of genera) and/or disk-florets, some unisexual, bisexual or sterile; petals fused.

Ovary inferior; style bifid.

Fruit achene, sometimes with wings or hairs or scales at apex.

Confused with:

Acanthaceae

Leaves opposite, often with cystoliths; flowers not in heads; ovary superior.

Habit of *Chromolaena odorata*

Mikania scandens

Asteraceae cont.

© R. de Kok

Habit of *Vernonia* sp.

Major genera:	
Adenostemma	Herbs; lower leaves opposite and petiolate; upper ones alternate and sessile; ray-florets absent; fruit without a tuft of hairs.
Ageratum	Herbs; lower leaves opposite; upper ones alternate; ray-florets absent; when dried with a sweet scent; fruit without a tuft of hairs, sometimes with scales.
Bidens	Herbs; leaves opposite, sometimes compound; ray-florets sometimes absent; fruit bristly.
Blumea	Herbs, shrubs and climbers; leaves alternate; ray-florets present; fruit with a tuft of hairs.
Crassocephalum	Herbs; leaves alternate; inflorescence nodding at anthesis; ray-florets absent; fruit with a tuft of hairs.
Chromolaena	Herbs; leaves usually opposite; ray-florets present; fruit with a tuft of hairs.
Dichrocephala	Herbs, rooting at base; leaves alternate; ray-florets present; fruit without a tuft of hairs.
Emilia	Herbs with ribbed stems; leaves alternate; ray-florets absent; fruit with a tuft of hairs.
Olearia	Trees, shrubs or herbs; leaves alternate to opposite, stellate hairs sometimes present; ray-florets usually present; fruit with bristles.
Senecio	Trees, shrubs or herbs; leaves alternate, with stellate hairs; ray-florets sometimes absent; fruit with a tuft of hairs.
Vernonia	Trees, shrubs, lianas or herbs; leaves alternate, lower surface often with glands; ray-florets absent; fruit with a tuft of hairs.
Wedelia	Herbs, often climbing; leaves opposite (upper ones sometimes alternate); ray-florets present; fruit without a tuft of hairs.

Begoniaceae

Field characters:

Herbs; stem succulent, sometimes woody at base; leaves alternate, simple, usually asymmetrical; stipules present; flowers unisexual, stamens many; ovary inferior, 3-locular.

Description:

Habit herbs, monoecious; stem succulent, sometimes woody at base.
Sap absent.
Stipules present.
Leaves alternate, simple, margins not entire, usually asymmetrical, often succulent.
Flowers unisexual, sepals and petals free, stamens many in male flowers.
Ovary inferior, 3-locular.
Fruit a capsule, often 3 winged.

Confused with:

Balsaminaceae
Leaves symmetric; stipules absent.
Gesneriaceae
Leaves opposite; stipules absent.
Urticaceae
Cystoliths present in the leaves; flowers small.

© RBG Kew

Habit of *Begonia* sp.

© J.Gregson

Female flower of *Begonia gueritziana*

Genus:

Begonia	As for the family.

Bignoniaceae

Field characters:

Trees or climbers; leaves opposite, compound; stipules absent; flowers showy, tubular; fruit capsules or pods; seeds many, winged.

Description:

Habit trees, shrubs or lianas.

Sap absent.

Stipules absent.

Leaves opposite or whorled, usually 1–4 pinnate compound or rarely unifoliate, margins entire.

Flowers bisexual, 5-merous, tubular to funnel-shaped corollas, showy; stamens 4–5.

Ovary superior, 2-celled.

Fruit a capsule or pod, divided lengthways; seeds many, flat, often with papery transparent wings.

Confused with:

Leguminosae (Caesalpinioideae, Mimosoideae and Papilionoideae)
Trees, lianas or herbs; leaves usually alternate; stipules usually present; petals free; stamens 10.

Gesneriaceae
Usually herbs; leaves simple; fruit a capsule or an indehiscent berry; seeds many, tiny.

Acanthaceae
Usually herbs, stems with swollen nodes (shrunken when dried); leaves simple, usually with cystoliths; fruit usually with hooks inside; seeds few.

Lamiaceae
Trees, lianas or herbs; leaves simple or palmate, seldom pinnate; fruit usually a drupe; seeds seldom with wings.

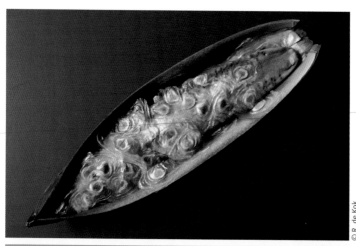

Capsule, divided lengthways, often with papery transparent wing

© R. de Kok

Genera:

Deplanchea	Trees; leaves whorled, simple, with 2 large glands at base of upper surface; capsule flat in cross-section.
Dolichandrone	Trees; leaves pinnate, turning black on drying, domatia present; capsule flat in cross-section; seeds with corky ridges.
Nyctocalos	Lianas; leaves trifoliate, without tendrils; corolla salver-shaped, with a long and narrow cylindrical tube; capsule flat in cross-section.
Oroxylum	Trees; leaves 2–4 pinnate, lower leaf surface sometimes with glands at the base; flowers fleshy; capsule flat in cross-section.
Radermachera	Trees; leaves 1–2(–3) pinnate, lower leaf surface always with glands; flowers not fleshy; capsule cylindrical in cross-section.

Burseraceae

Field characters:

Large trees; leaves alternate, compound, usually imparipinnate; leaflets often opposite, distinct petiolules; usually stipules absent (sometimes pseudo-stipules present); stamens free; ovary superior.

Description:

Habit often large trees, sometimes shrubs.

Sap resin present.

Stipules absent, pseudo-stipules sometimes present.

Leaves alternate, pinnately compound, rarely unifoliate or trifoliolate, often imparipinnate; leaflets opposite, margins entire or toothed, with strong resinous smell when crushed; petiolules usually with distinct basal and apical swellings.

Flowers small, usually unisexual, 3–5-merous, disk present; stamens usually free, equal in number or twice as many as petals, glabrous.

Ovary superior, 2 (rarely 1) ovules per cell.

Fruit a drupe or very rarely a capsule, often 3-sided; single-seeded.

Confused with:

Anacardiaceae

White sap that turns black on exposure; leaves usually simple; leaflets opposite to alternate, entire; petiolules short; 1 ovule per cell.

Meliaceae

Not resiniferous; leaves without odour; leaflets alternate or sub-opposite, usually entire, petiolules short; stamens united in a tube, fruit usually globose.

Sabiaceae

Not resiniferous; leaves without odour; leaflets opposite, entire to dentate; petiolules short; petals opposite the sepals; 1 ovule per cell.

Sapindaceae

Not resiniferous; leaves paripinnate, without odour; leaflets alternate or opposite, entire to dentate; petiolules short; stamens hairy; 1 ovule per cell.

Pseudo-stipules of *Canarium denticulatum*

© J. Gregson

Petiolules of leaflets with distinct swellings of *Dacryodes* sp.

© J. Dransfield

Cross-section of dried fruit of *Canarium indicum*

© R. de Kok

Burseraceae cont.

© J. Gregson

Habit of *Canarium denticulatum*

Genera:	
Canarium	Leaflet margins entire to toothed; pseudo-stipules sometimes present; flowers 3-merous; fruit a drupe, hairy or glabrous, usually blue-black when mature; seeds 3, 1 seed much bigger than the other 2; fruiting calyx enlarged.
Dacryodes	Leaflet margins entire; pseudo-stipules sometimes present; flowers 3-merous; fruit a drupe, glabrous, wrinkled when dried; stigma not off-centre; single-seeded.
Santiria	Leaflet margins entire, pseudo-stipules absent; flowers 3-merous; fruit brightly coloured drupes, smooth when dry; stigma distinctly off-centre; single-seeded.
Triomma	Leaflet margins entire, pseudo-stipules absent; flowers 4–5-merous; fruit a triangular capsule; seeds 3, winged.

Capparidaceae

Flowers of *Crateva magna*

Field characters:

Leaves spirally arranged; stipules usually present; flowers regular or nearly so; ovary superior, sitting on a gynophore.

Description:

Habit trees, shrubs, climbers or herbs with spines.

Sap absent.

Stipules usually present, sometimes thorn-like.

Leaves spirally arranged, margins entire, simple to 3-foliolate or palmate, often with pellucid dots.

Flowers regular or almost so, (0–)4(–6)-merous; petals free, seldom fused; stamens 4–many.

Ovary superior, 1(–3)-locular.

Fruit a berry or a woody capsule, sitting on a gynophore; seeds 1–many.

Gynophore of *Capparis* sp.

Confused with:

Cruciferae

Herbaceous; flowers regular; ovary 2-locular.

Crateva magna

Genera:

Capparis	Shrubs or climbers; leaves simple; stipules thorn-like, usually present; stamens many, free; fruit a berry.
Cleome	Herbs; leaves palmately lobed; sometimes with spines; stamens 6–many; fruit a capsule.
Crateva	Trees; leaves palmately lobed to 3-foliolate; stipules small; stamens 8–30; fruit a berry.
Stixis	Shrubs or climbers unarmed; leaves simple; stamens 20–100; fruit a berry, single-seeded.

Caprifoliaceae

Field characters:

Leaves opposite, penninerved; inflorescences terminal, cymose; corolla white, fused; ovary inferior, 2–5-locular.

Description:

Habit trees, shrubs, lianas or sometimes herbs; when woody stem and twigs often with thick pith.

Sap absent.

Stipules usually absent.

Leaves opposite to alternate, simple, rarely compound, margins often dentate, penninerved.

Flowers (4–)5-merous; calyx fused, (4–)5-lobed; corolla fused, sometimes 2-lipped; stamens 4–5.

Ovary inferior, 2–5-locular.

Fruit a drupe or berry, sometimes an achene or a capsule, often compressed and ridged.

Confused with:

Rubiaceae

Leaves simple, entire; interpetiolar stipules present.

© R. de Kok

Sambucus sp.

© R. de Kok

Viburnum sp.

Genera:	
Sambucus	Leaves pinnately or bipinnate; fruit a 3–5-seeded berry.
Viburnum	Leaves simple; fruit a single-seeded drupe.

Celastraceae

Field characters:

Trees, shrubs and lianas; leaves usually opposite, simple; stipules small; flowers small, conspicuous disk often present; stamens the same in number as and opposite the sepals; ovary superior.

Description:

Habit trees, shrubs or lianas.

Sap absent.

Stipules absent or small and falling off early.

Leaves opposite to sub-opposite or alternate, simple; margins often dentate, rarely entire; penninerved, dried leaves often pale grey-green, sometimes with black glandular dots.

Flowers bisexual or unisexual, regular, 4–5-merous; calyx persistent; petals free; disk present; stamens same in number or fewer than the sepals, opposite the sepals.

Ovary superior, 2–5-locular, stigma narrow to broad.

Fruit usually a capsule with arillate or sometimes winged seeds; sometimes a drupe or a berry.

Confused with:

Aquifoliaceae
Leaves alternate; disk absent; stigma broad, sessile; fruit drupaceous with 3 or more pyrenes.

Flacourtiaceae
Leaves spirally arranged; flowers often bisexual, with many stamens; ovary 1-locular.

Rhamnaceae
Leaves rarely opposite, venation sclariform; stamens opposite the petals.

© J. Dransfield

Fruit of *Euonymus* sp.

Genera:

Cassine	Trees or shrubs; leaves usually (sub-)opposite; fruit an indehiscent capsule.
Celastrus	Liana; leaves spirally arranged, margins serrate or dentate; fruit a 3-locular capsule; seeds 1–6, enveloped with a fleshy crimson aril.
Euonymus	Trees or shrubs; leaves usually (sub-)opposite, when carefully broken in two, the parts are still connected by white threads; fruit a loculicidal capsule; seeds with an aril.
Kokoona	Trees or shrubs; leaves usually (sub-)opposite; stigma broad; fruit a 3-locular capsule; seeds with a wing attached to the apex.
Loeseneriella	Liana; leaves usually opposite; fruit capsular with 3-follicles; seeds winged.
Lophopetalum	Trees with pneumatophores; leaves usually (sub-)opposite; fruit a 3-locular capsule; seeds winged.
Microtropis	Trees or shrubs; leaves usually (sub-)opposite; fruit a capsule with a single split along one side; seed usually 1, enveloped with an aril.
Reissantia	Liana; leaves usually opposite; fruit a 3-locular capsule; seeds winged.
Salacia	Liana; leaves usually opposite; fruit drupaceous; seed 1–several, embedded in pulp.
Siphonodon	Trees or shrubs, leaves usually spirally arranged, margins toothed; ovary many-celled; fruit drupaceous, hard; seed 1, flat.

Clusiaceae (Guttiferae)

Field characters:

Trees or shrubs, usually with yellow (rarely red) sap; leaves opposite, simple, margins entire; stipules absent; ovary superior.

Description:

Habit trees or shrubs.

Sap yellow from cut surfaces, rarely red.

Stipules absent.

Leaves opposite, usually decussate, simple, margins entire, often with many lateral veins, with resinous gland dots or lines; petioles with basal excavations (pouches).

Flowers bisexual or unisexual, regular, 4–5-merous; sepals and petals free, sepals imbricate, petals contorted; stamens many, often in bundles.

Ovary superior; stigma large, peltate.

Fruit a drupe, nut, capsule or a berry; 1–many seeded, often arillate.

Confused with:

Apocynaceae
White sap present; corolla tubular; fruit paired.

Celastraceae
Sap absent; leaves often dentate; stamens 4–5.

Loganiaceae
Sap absent; petals fused; stamens 4–5.

Theaceae
Sap absent; leaves spirally arranged, margins usually dentate.

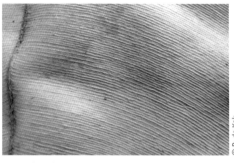

Detail of the lateral veins in the leaves of *Calophyllum* sp.

Flower of *Hypericum laschenaultii*   Flowers of *Garcinia* sp.   Leaf pouches of *Garcinia maingayi*

Genera:

Cratoxylum	Trees with red sap; leaves with dots; flowers bisexual; fruit a dehiscent capsule; seeds winged.
Garcinia	Usually trees, sap yellow or red; leaves with fine resin ducts crossing the veins; flowers unisexual; fruit a fleshy to woody berry.
Calophyllum	Trees with red or yellow sap; leaves with very dense parallel venation; flowers sometimes unisexual; fruit a drupe.
Mammea	Trees with yellow sap; leaves with dots; flowers unisexual; fruit a drupe.
Mesua	Trees or shrubs with yellow sap; flowers bisexual; fruit a nut.

Commelinaceae

Field characters:

Herbs, often with rather succulent leaves or stems; leaves with a closed basal sheath, often longitudinally striped or ridged; ligule absent; flowers with (at least inner) tepals rapidly deliquescent (in a few hours), anthers often hairy.

Description:

Herbs often with rather succulent leaves or stems; roots fibrous or tuberous.

Sap absent, but may have clear juice when cut.

Stipules absent.

Leaves spirally arranged, with a basal sheath, often longitudinally striped or ridged; ligule absent; blade simple, margins entire; often pseudo-petiolate; hairs often present, at least on leaf sheaths.

Flowers tepals in 2 whorls, outer 3-sepaloid or petaloid, inner 3-petaloid, at least inner whorl withering; stamens usually 6, often hairy.

Ovary superior, 2- or 3-locular.

Fruit usually a dehiscent capsule, rarely indehiscent or a berry.

Confused with:

Zingiberaceae
Leaves alternate, sheath open; ligule present.

Marantaceae
Leaf blade apex offset relative to midvein; petiole swollen apically.

Orchidaceae
Stamens and style forming a column; roots covered in papery or velvet sheath.

Poaceae
Ligule present; flowers without a petaloid whorl; perianth of bristles or scales.

© P. Wilkin

Inflorescence of *Amischotolype glabrata*

© RBG Kew

Flowers of *Commelina benghalensis*

Major genera:

Amischotolype	Creeping herbs; lamina petiolate; inflorescence opposite the leaf base and piercing the leaf-sheath; flowers actinomorphic, stamens 6.
Belosynapsis	Herbs, always prostrate, sometimes epiphytic; inflorescence terminal or axillary; flowers actinomorphic, stamens 6.
Commelina	Herbs; leaves usually distichous; inflorescence terminal or leaf-opposite, with a folded spathe; flowers strongly zygomorphic, stamens 3.
Pollia	Creeping herbs; lamina petiolate; inflorescence terminal; flowers actinomorphic or almost so, stamens 6; fruit berry-like, usually metallic blue.
Tricarpelema	Herbs; leaves spirally arranged or distichous, usually petiolate; inflorescence terminal or axillary; flowers zygomorphic, stamens 6.

Connaraceae

Field characters:

Trees, shrubs or usually climbers; leaves spirally arranged, compound; stipules absent; sepals and petals free; ovary superior; fruit a single-seeded capsule; seeds arillate.

Description:

Habit usually climbers, sometimes shrubs or small trees.

Sap absent.

Stipules absent.

Leaves alternate, 1–4 pinnately compound, rarely unifoliolate, usually imparipinnate; margins entire; penninerved, pellucid dots sometimes present; base of petioles swollen.

Flowers bisexual, 4–5-merous; sepals and petal free or joined at base; stamens usually 10.

Ovary superior, often with 1–5-free carpels.

Fruit single-seeded capsules, sometimes several free capsules per fruit; seeds arillate.

Confused with:

Legumes (Caesalpinioideae, Mimosoideae and Papilionoideae)
Stipules present; sepals connate; ovules usually in a row; fruit mostly with more than 1 seed.

Fruit and seed with aril of *Ellipanthus beccarii* var. *peltatus*

Fruit of *Agelaea borneensis*

© S. Marsh

© S. Marsh

Ellipanthus beccarii var. *peltatus*

Genera:	
Agelaea	Lianas or shrubs; leaves trifoliolate; fruit red, warty.
Cnestis	Lianas or shrubs; leaves pinnate, leaflets >10 pairs, without emarginate apex; fruit brown, velvety.
Connarus	Trees or lianas and shrubs; leaves palmate to pinnate, pellucid dots present, without emarginate apex; fruit glabrous.
Ellipanthus	Trees; leaves unifoliate; petiole apex and base swollen; fruit yellowish, velvety.
Rourea	Lianas or shrubs; leaves imparipinnate, leaflets <10 pairs, without emarginate apex; fruit smooth, glabrous.
Roureopsis	Lianas or shrubs; leaves imparipinnate; leaflets >(1–)10 pairs, with emarginated apex; fruit red, smooth, glabrous.

Convolvulaceae

Field characters:

Mostly vines; leaves spirally arranged, simple; tendrils absent; stipules absent; sepals free; corolla tubular; ovary superior.

Description:

Habit twining climbers or creeping herbs without tendrils; rarely shrubs, small trees or parasitic climbers.

Sap rarely present, when present watery to white.

Stipules absent.

Leaves spirally arranged, simple, very rarely compound, margins usually entire; venation pinnate to palminerved; hairs simple, sometimes stellate or T-hairs.

Flowers bisexual, regular, usually 5-merous, sometimes subtended by a pair of bracts, these sometimes enlarged; sepals free and overlapping; corolla tubular, usually large; stamens 5, fused to the corolla tube.

Ovary superior, 2-celled.

Fruit usually a dehiscent capsule, sometimes a berry; often with persistent scarious to woody sepals; seeds sometimes hairy, with a hard shiny testa.

Confused with:

Cucurbitaceae
Tendrils at 90° relative to leaf; ovary inferior.

Passifloraceae
Tendrils in leaf axils; stipules present; ovary superior.

Vitaceae
Tendrils opposite the leaves; ovary superior.

Lauraceae, *Cassytha*
Flowers 3-merous; fruit a berry.

© R. de Kok

Habit of *Ipomoea* sp.

© T. Utteridge

Flowers of *Erycibe* sp.

© M. Coode

Cuscuta sp.

© R. de Kok

Flower of *Ipomoea cairica*

Genera:

Argyreia	Mostly woody climbers; styles present; fruit a berry.
Cuscuta	Parasitic climber without leaves; fruit a dry or fleshy capsule.
Erycibe	Shrubs, small trees or woody climbers; sometimes with T- or stellate hairs; petals lobed, bifid; styles absent; fruit a berry.
Ipomoea	Herbs and shrubs, usually climbers or prostate; sometimes with stellate hairs; sometimes with a watery to milky sap; styles present; fruit a capsule.
Jacquemontia	Herbaceous or woody climber; leaves and stems often with stellate hairs; styles present; fruit a capsule.
Merremia	Herbs or shrubs, usually climbing; styles present; fruit a capsule.

Costaceae

Field characters:

Rhizomatous herbs, not aromatic; leaves spirally arranged, with a closed sheath; ligule forming a ring above the pseudo-petiole; tepals fused into a tube, at least at base; 1 fertile stamen; ovary inferior.

Description:

Habit large terrestrial herbs, not aromatic; stems often semi-woody, branched; leaves often spirally arranged; rhizomatous.

Sap absent.

Stipules absent.

Leaves spirally arranged, simple, with a closed sheath; ligule forming a ring above the pseudo-petiole; margins entire; hairs simple or branched.

Flowers bisexual; tepals fused into a tube, at least at base; outer whorl (calyx) usually 3-toothed or lobed; inner whorl 3-lobed; only 1 fertile stamen present, lateral staminodes usually present.

Ovary inferior.

Fruit a capsule, (2–)3-locular; seeds numerous.

Confused with:

Zingiberaceae
Leaves alternate; leaf sheath open; ligule a small leaf-like structure at the base of the leaf blade.
Marantaceae
Petiole swollen apically; stamens petaloid.
Orchidaceae
Stamens and style forming a column; roots covered in papery or velvet sheath.

© J. Gregson

Inflorescence of *Costus* sp.

Habit of *Costus* sp.

Genus:

Costus	As for the family.

Cucurbitaceae

Field characters:

Climbers with tendrils at 90° relative to the leaf; flowers unisexual, always white and/or yellow; ovary inferior; fruit usually a fleshy berry.

Description:

Habit climbers or creepers, to lianas, tendrils at 90° relative to the leaves.

Sap absent.

Stipules absent.

Leaves spirally arranged, simple, usually deeply lobed, palmately veined; margins toothed to entire; hairs scabrid or hispid, often glandular.

Flowers unisexual, 3- or 5(–7)-merous, yellow or white, very rarely red or greenish; corolla tubular; stamens 5 or reduced to 3, free or fused, often much bent and coiled.

Ovary inferior, style 1–3; stigmas usually large, bilobed.

Fruit usually a fleshy berry, sometimes an explosive or dehiscent capsule; seeds 1 to usually many, flattened; sometimes winged.

Confused with:

Convolvulaceae
Tendrils absent; flowers bisexual; ovary superior.

Passifloraceae
Tendrils in leaf axils; stipules present; flower bisexual; ovary superior.

Vitaceae
Tendrils opposite the leaves; flowers bisexual, small; ovary superior.

© R. de Kok

Tendril at 90° relative to the leaves of *Luffa acutangula*

© R. de Kok

Fruit of *Neoalsomitra* sp.

Cucurbitaceae cont.

© R. de Kok

Flower of *Momordica* sp.

Major genera:	
Alsomitra	Large climber; leaves entire; corolla lobes entire; stamens 3; fruit capsular; seeds winged.
Gymnopetalum	Climber or creeping herb; leaves simple, often lobed; corolla lobes entire; stamens 3; fruit a berry; seeds not winged.
Hodgsonia	Tall woody climber; leaves palmately lobed; corolla lobes bifid; stamens 3; fruit woody; seeds 6.
Momordica	Climber; leaves simple, often lobed; corolla lobes entire, stamens 3; fruit a berry, with soft spines, warts or ridges present.
Neoalsomitra	Small- or medium-sized herbaceous or woody climbers, leaves simple to palmately lobed; corolla lobes entire; stamens 5, free to fused; fruit a cylindrical capsule with a truncate apex; seeds winged.
Trichosanthes	Small to large climber; leaves simple to palmately lobed; corolla lobes fimbriate; stamens 3; fruit a smooth berry.

Cyperaceae

Field characters:

Herbs; stems solid, without joints; leaves often 3-ranked; flowers subtended by bracts (i.e. glumes), arranged in spikelets.

Description:

Habit herbs with underground rhizomes or stolons in perennial species.

Sap absent.

Stems (culms) often triangular in cross-section, solid, without nodes.

Leaves often 3-ranked, base of leaf sheathing the stem.

Inflorescence usually of spikelets arranged into panicles.

Flowers minute with/without a reduced perianth, stamens 3, utricle (perygynium) present in half of the species, lemma and palea absent.

Ovary usually superior, stigmas 2–3.

Fruit a single-seeded nutlet, 2–3 carpels.

Confused with:

Poaceae
Stems often hollow, with joints; leaves often 2-ranked and/or ligulate, each floret subtended by 2 bracts.

Juncaceae
Stems solid to hollow, without joints, leaves spirally arranged, flowers not subtended by a bract.

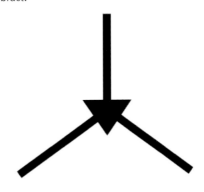

Stems solid, often triangular in cross-section, with leaves in 3-ranks

Cyperus sp.

© R. de Kok

Major genus:

Many genera of Cyperaceae, too numerous to list in this non-specialist field guide, are found in East Sabah.

Dipterocarpaceae

Field characters:

Trees; leaves alternate, simple, margins entire; stipules present; calyx persistent, often enlarged at fruiting into (0–)3–5 wings; corolla contorted in bud.

Description:

Habit trees, often large.

Sap resins sometimes present.

Stipules present, often large and protecting the terminal bud.

Leaves alternate, coriaceous, simple, margins entire; venation clearly visible, side veins parallel but not particularly close, sometimes with an intramarginal vein; hairs often stellate or tufted; sometimes with scales; petioles often swollen apically.

Flowers bisexual, 5-merous, often nodding, pointed in bud; calyx with equal or unequal wings; corolla conspicuously contorted with more-or-less free petals; stamens 5-merous.

Ovary superior, rarely semi-inferior.

Fruit a dry, indehiscent, single-seeded nut; usually winged with the persistent calyx lobes strongly accrescent surrounding the nut.

Confused with:

Hernandiaceae
Fruit with 2 wings (wings are enlarged calyx lobes on the apex of the fruit).

Anacardiaceae
White sap, which turns black on exposure.

Apocynaceae
White sap; leaves opposite; stellate hairs absent.

© T. Utteridge

© T. Utteridge

Contorted corolla of *Parashorea* sp.

Flowers of *Vatica rassak*

Fruit of *Shorea parvistipulata*

Fruit of *Vatica rassak*

Genera:	
Anisoptera	Leaves with intramarginal vein and with scales below; stipules linear and falling off early; petiole swollen apically; ovary (semi-)inferior; fruit wings unequal.
Cotylelobium	Leaves with intramarginal vein and without scales below; stipules falling off early; petiole not swollen apically; ovary superior; fruit wings unequal.
Dipterocarpus	Leaves with parallel straight side veins and no intramarginal vein; stipules large; petiole swollen apically; ovary (semi-)inferior; fruit wings unequal or absent.
Dryobalanops	Wood smelling of camphor; leaves with intramarginal vein and dense numerous parallel side veins, not silvery below; stipules linear and falling off early; petiole slender; ovary superior; fruit wings equal to sub-equal.
Hopea	Leaves with arching side veins and no intramarginal vein, rarely velutinous below; stipules linear and falling off early; petiole rarely swollen apically; ovary superior; fruit wings unequal.
Parashorea	Leaves with parallel straight side veins and no intramarginal vein, silvery below; stipules usually small and falling off early; petiole slightly swollen apically; ovary (semi-)inferior; fruit wings unequal.
Shorea	Leaves with arching side veins and no intramarginal vein, rarely velutinous below; stipules large and falling off early; petiole rarely swollen apically; ovary superior; fruit wings (un)equal or absent.
Upuna	Leaves with arching side veins and no intramarginal vein, whitish to brown, velutinous below; stipules linear and persisting; petiole swollen apically; ovary superior to (semi-)inferior; fruit wings unequal.
Vatica	Leaves with arching side veins and no intramarginal vein, rarely velutinous below; stipules variable; petiole rarely swollen apically; ovary superior to (semi-)inferior; fruit wings (un)equal or absent.

Durionaceae (Bombacaceae)

Field characters:

Trees; leaves simple, usually oblong-elliptic, nerves pinnate; lower surface with many scales or stellate hairs; fruit spiny; seeds arillate; ovary semi-inferior.

Description:

Habit trees.

Sap absent.

Stipules present, but often falling off early.

Leaves alternate, simple, usually oblong-elliptic; margins entire, pinnately nerved; lower surface with scales or stellate hairs.

Flowers bisexual, 5-merous, showy; stamens numerous, connate at base and with corolla; epicalyx often present.

Ovary superior.

Fruit a spiny capsule, inner surface sometimes with irritant hairs; seeds arillate.

Confused with:

Malvaceae

Shrubs or small trees; stamens fused in a tube and attached to the corolla.

Sterculiaceae

Shrubs or trees; stamens 5–20, free or fused into a terminated head or central bundle; fruit smooth; carpels divided.

Sparrmanniaceae

Shrubs or small trees; leaves often palminerved; stamens free or rarely united in bundle(s).

© R. de Kok

Outer surface of the fruit of *Durio* sp.

© J. Dransfield

© R. de Kok

Detail of leaf with scales of *Durio* sp.

Inside of *Durio* sp. showing the fleshy aril

Genera:

Durio	Leaf acute to acuminate; fruit globose, dehiscent from top to base before or after falling from tree, inner fruit wall glabrous; seeds with a fleshy aril.
Neesia	Leaf tip rounded to notched; fruit 5-angled, dehiscent from top to base before falling from tree, inner fruit wall with irritant hairs; seeds without a fleshy aril.

Ebenaceae

Field characters:

Trees with black twigs; leaves alternate, simple, margins entire, often with glands on the lower surface; calyx lobed, persistent, enlarged, and often reflexed in fruit.

Description:

Habit trees; often with black bark.

Sap absent.

Stipules absent.

Leaves usually alternate, rarely opposite or spirally arranged; simple, margins entire; T-hairs present; often with glands near the base or mid-vein on the lower surfaces.

Inflorescences axillary, many-flowered (usually male) or reduced to a single flowers (usually female).

Flowers unisexual, dioecious, regular, 3–5(– 8)-merous, often white or cream; sepals fused at least at the base, lobes valvate or imbricate; petals fused at least at the base (corolla tubular, campanulate or urceolate), often equal in number to sepals, may be forked, fringed or hairy; stamens usually attached to the base of the petals.

Ovary superior, rarely inferior, sessile, with as many locules as there are sepals and petals; stigmas often lobed.

Fruit a berry, usually indehiscent; mesocarp often fleshy, with a stony inner part; subtended by the enlarged and persistent calyx.

Confused with:

Annonaceae
Twigs with longitudinal ridges; rays in a cross-section in the wood; flowers bisexual, parts free.

Euphorbiaceae
Sap sometimes present; margins often dentate; stipules often present.

Symplocaceae
Leaves serrate, turning yellow when drying; ovary inferior.

Sapotaceae
White sap present; flowers bisexual, parts free.

Fruit with recurved calyx lobes of *Diospyros kaki*

Flower of *Diospyros kaki*

Leaf-glands of *Diospyros* sp.

© M. Coode

© M. Coode

© Tim Utteridge

Genus:

Diospyros   As for the family.

Elaeocarpaceae

Field characters:

Trees and large shrubs; leaves usually spirally arranged, simple; stipules present; flowers regular; sepals free, valvate; petals usually free, margins usually fimbriate.

Description:

Habit trees and large shrubs.

Sap absent.

Stipules present, often falling off.

Leaves alternate to opposite, rarely whorled; simple, margins often serrate, stellate hairs absent, domatia often present; petiole often bipulvinate.

Flowers usually bisexual, regular, 4–6-merous; sepals free, valvate; petals usually free, margins usually fimbriate; disk present; stamens usually >10; anthers basifixed, dehiscing from terminal pores or lateral slits.

Ovary superior, 2–7-locular, style 1.

Fruit a drupe or a capsule; seeds sometimes sculptured.

Confused with:

Simaroubaceae
Leaves simple or pinnate; stipules leaving an annular scar; often with free carpels in each flower.

Sparrmanniaceae
Leaves often 3-veined from the base, with stellate hairs; anthers dorsifixed.

Sterculiaceae
Leaves with stellate hairs; flowers usually with 1 perianth whorl; stamens usually 5, anthers dorsifixed; carpels divided.

© T. Utteridge

Fruit of *Elaeocarpus* sp.

© R. de Kok

Flowers of *Elaeocarpus* sp.

© RBG Kew

Habit of *Elaeocarpus* sp.

Genera:

Elaeocarpus	Petals margins fimbriate; stamens inserted between the disk and ovary; fruit drupaceous, often blue or purple-black.
Sloanea	Petioles swollen apically; petals margins sometimes fimbriate; stamens inserted on the disk; fruit capsular, sometimes spiny.

Ericaceae

Trees, shrubs, often epiphytic; leaves spirally arranged; stipules absent; corolla urceolate or tubular; ovary superior to inferior; stamens 10, dehiscing by terminal pores.

Description:

Habit trees, shrubs, often epiphytic, rarely lianas.

Sap absent.

Stipules absent.

Leaves usually spirally arranged, rarely pseudoverticillate or opposite; simple, margins entire to toothed; young leaves often red; venation pinnate sometimes with 3 or more secondary veins from the base; hairs when present, usually simple, often stellate hairs or scales.

Inflorescences terminal or axillary, sometimes in bracteate racemes.

Flowers bisexual, usually 5-merous; corolla urceolate or tubular; stamens 10, dehiscing with terminal pores, anthers or filaments with appendages.

Ovary superior to inferior.

Fruit a capsule, berry or drupe; calyx persistent; seeds numerous.

Confused with:

Epacridaceae

Leaves with close parallel veins clearly visible on the lower surface.

Myrsinaceae

Leaves with dark dots; petals connate at base, contorted stamens.

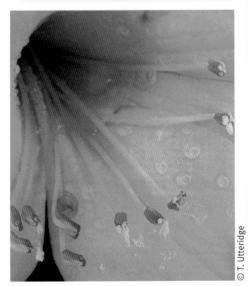

Anther-pores of *Rhododendron* sp.

Vaccinium sp.

© T. Utteridge

© T. Utteridge

Ericaceae cont.

Rhododendron sp.

Genera:	
Costera	Shrubs; leaves entire; flowers solitary or in few fascicles; fruit a berry; ovary inferior.
Diplycosia	Shrubs, mainly epiphytic; leaves usually entire; flowers in bracteate racemes; fruit a capsule; ovary superior.
Gaultheria	Shrubs, rarely epiphytic; leaves mostly serrate; flowers in bracteate racemes; fruit a septicidal capsule; ovary superior.
Rhododendron	Shrubs or trees; leaves entire to serrate, usually conspicuous with stellate hairs or scales; fruit a septicidal capsule; ovary superior.
Vaccinium	Trees or shrubs, often epiphytic; leaves entire to serrate; flowers in bracteate racemes, rarely solitary; fruit a berry; ovary inferior.

© S. Marsh

Euphorbiaceae s.l.

Contributed by Chill Chalen

Field characters:

Usually trees or shrubs, white or red sap sometimes present; leaves alternate, simple; glands on petiole or blades sometimes present; stipules present; flowers unisexual; fruit often 3-lobed; ovary superior.

Description:

Habit trees and shrubs, sometimes herbs or climbers, rarely spiny.

Sap absent or white, reddish or clear.

Stipules present, sometimes foliaceous.

Leaves alternate (spirally or distichously arranged), rarely opposite, simple or palmately compound; margins entire or not; venation pinnate or palmate; often with glands at the base of leaves; sometimes with domatia; petioles often long and with swellings at the apex and base; hairs simple or branched, stellate, T-shaped or with scales.

Flowers unisexual (plants dioecious or monoecious), small, usually yellowish green, rarely with coloured petals or bracts; usually 5-merous, petals and disc present or absent.

Ovary superior, usually with a single 3-armed style, (1–)3(–20)-locular, placentation apical axial.

Fruit mostly a capsule opening explosively on drying, sometimes sculptured; sometimes drupes or berries, rarely samaras; seeds often carunculate, sometimes arillate, otherwise without appendages.

Confused with:

Buxaceae
Stipules absent.

Celastraceae
Flowers often bisexual, usually with a prominent floral disk.

Daphniphyllaceae
Stipules absent; flowers with a bilobed style; fruit single-seeded drupe.

Flacourtiaceae
Stipules often minute or absent; flowers often bisexual; ovary 1-locular, placentation parietal.

Theaceae
Sap and stipules absent; flowers bisexual, usually big, usually solitary.

Ulmaceae
Perianth in a single whorl, stamens opposite the tepals; fruit indehiscent, with 2 styles.

Urticaceae
Cystoliths present in leaves; ovary 1-locular; fruit single-seeded.

Fruit of *Antidesma* sp.

Fruit of *Baccaurea* sp.

Euphorbiaceae s.l. cont.

Glochidion sp.

Fruit of *Glochidion* sp.

Mallotus sp.

Fruit of *Mallotus* sp.

Suregada sp.

Euphorbiaceae s.l. cont.

Major genera:	
Antidesma	Trees and shrubs; inflorescence an axillary raceme; fruit a drupe, often latterly compressed, sculptured when dry.
Aporusa	Small trees, sap absent, simple hairs present; inflorescence looks like a spike.
Baccaurea	Trees and shrubs with *Terminalia*-type branching pattern, sometimes with stellate hairs; inflorescence axillary and cauline; fruit vary from berries to dehiscing fleshy capsules, 2–4 locules.
Bridelia	Trees and shrubs, simple hairs present; male and female flowers with petals and disk; fruit usually an indehiscent berry or drupe, 2–3-locular.
Cleistanthus	Trees and shrubs, simple hairs present; leaf margins entire; inflorescence a fascicle; flowers symmetric, 5-merous; fruit a 3-locular smooth capsule.
Croton	Trees to herbs, often with red sap, often with scales and stellate hairs; leaves usually with 2 trumpet-shaped glands at base; fruit a dehiscent 3-locular capsule.
Drypetes	Trees and shrubs; leaves usually with asymmetric base; flowers axillary or cauline, petals absent; fruit 1–4-locular fleshy drupe.
Glochidion	Trees and shrubs, simple hairs present; flowers have stamens or styles fused in a column; fruit 3–25-locular, 2 seeds per locule.
Macaranga	Trees or shrubs, sometimes with red sap, simple hairs present; leaves often with granular glands on lower surface; fruit usually 3-locular, sometimes with spines.
Mallotus	Trees or shrubs, sometimes with red sap, stellate hairs sometimes present; leaves often with granular glands on lower surface; fruit usually 3-locular, sometimes with spines.
Phyllanthus	Trees to herbs, no sap or glands; main stem with reduced scales and normal leaves on side stems, the whole twig often looks like a compound leaf; flowers and fruit axillary on lateral shoots.
Trigonostemon	Small tree to shrubs; red sap present; male and female flowers with 5-petals, often brightly coloured; fruit a 3-locular capsule, can be smooth and warty.
Suregada	Small trees and shrubs; leaves drying pale green; inflorescence opposite a leaf; flowers with no petals; fruit usually a dehisced, 3-locular capsule or drupaceous.

Fagaceae

Field characters:

Trees; leaves usually alternate, simple; stipules present, falling off early; inflorescences arranged in catkins, heads or spikes; ovary inferior; fruit surrounded by a woody cupule or cup-like involucre.

Description:

Habit trees, rarely shrubs.

Sap absent.

Stipules present, but soon falling off; sometimes interpetiolar.

Leaves alternate, rarely opposite or whorled, simple, margins often entire, venation pinnate.

Inflorescences male erect or pendulous; female and mixed inflorescences erect.

Flowers small, unisexual, monoecious or dioecious, without petals, regular; male flowers with 4 to many stamens, filaments free, slender and distinct; female flowers often surrounded by basal involucre.

Ovary inferior, 3–9 styles and locules.

Fruit single-seeded nuts surrounded by a woody cupule or cup-like involucre.

Confused with:

Lauraceae
Stipules absent; ovary superior, sometimes seemingly inferior; fruit a fleshy cupule if present.

Myricaceae
Ovary superior; fruit a drupe

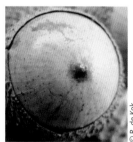

Acorn umbo of *Lithocarpus* sp.

© R. de Kok

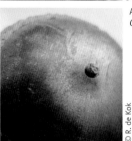

Acorn umbo of *Quercus* sp.

© R. de Kok

Fruit of *Castanopsis* sp.

© R. de Kok

© M. Coode

Lithocarpus sp.

Genera:	
Castanopsis	Trees; leaves spirally arranged, margins entire to rarely serrate above the middle; petioles slender and distinctly thickened at base; male flowers with 10–12 stamens; cupule always with vertical sutures and often lobed and dehiscing, usually spiny, often containing several fruit; fruit rounded or angular in cross-section.
Lithocarpus	Trees, rarely shrubs; leaves spirally arranged, margins entire; petiole thick throughout; male inflorescences erect, male flowers with 10–12 stamens; cupule never with vertical sutures and never lobed, in clusters or solitary, often almost completely enclosing the single fruit (acorn), fruit rounded in cross-section, acorn umbo not ringed.
Quercus	Trees; leaves spirally arranged, margins often serrate; petiole slender and distinctly thickened at base; male inflorescences pendulous, male flowers usually with <7 stamens; cupule never lobed nor with vertical sutures and always solitary, cup-shaped or very shallow with only 1 fruit (acorn); fruit rounded in cross-section, acorn umbo ringed.
Trigonobalanus	Trees; leaves in whorls of 3, margins serrate, stipules interpetiolar; male inflorescences erect, male flowers usually with <7 stamens; cupule always lobed, grouped into clusters along rachis, usually containing several fruit; fruit 3-angled in cross-section.

Flacourtiaceae

Contributed by Sue Zmarzty

Field characters:

Trees or shrubs; leaves usually spirally arranged, simple; stipules absent or small; ovary 1-locular.

Description:

Habit trees, shrubs or climbers.

Sap absent.

Stipules absent or small, rarely foliaceous.

Leaves spirally arranged, rarely opposite, simple, margins entire to serrate, domatia sometimes present, glandular; dots and dashes often present, in particular around the midvein and along margins; petioles often thickened at base and/or apex.

Flowers unisexual or bisexual, often small, regular; sepals 3–12; petals 3–20 or absent; stamens 5–many, mostly free, rarely in a column; disk often present.

Ovary superior to rarely semi-inferior, mostly free; 1-locular or incompletely 3–10-locular; styles 1–10.

Fruit fleshy to dry berry or a capsule, rarely a drupe.

Confused with:

Euphorbiaceae

Usually with stipules; flowers unisexual; ovary usually 3-locular.

Celastraceae

Leaves often opposite, not glandular; stamens the same number as and opposite the sepals; ovary 2- or more locular.

Fruit of *Casearia* sp.

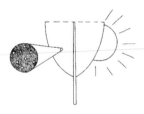

Dots and dashes are visible with a hand lens when a leaf is held against a strong light

© RBG Kew

Hydnocarpus sp.

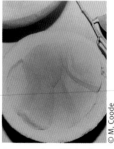

Cross-section of fruit of *Hydnocarpus* sp.

Major genera:

Casearia	Shrubs or trees; leaves with marginal glands, often with pellucid dots and dashes; flowers bisexual, hairy glands between stamens; fruit a 3-locular capsule.
Flacourtia	Trees and shrubs, usually armed; leaves with marginal glands; flowers bisexual or unisexual, stamens numerous; styles 5, persistent.
Homalium	Trees or shrubs; leaves with marginal glands; inflorescence usually a panicle or spike-like raceme; flowers bisexual, ovary semi-inferior.
Hydnocarpus	Trees and shrubs; leaves base sometimes unequal, without marginal glands; petiole swollen top and bottom; flowers unisexual; corolla with appendages on the inner surface.
Pangium	Trees with *Terminalia*-type branching; leaves without marginal glands, 3-veined; flowers unisexual; corolla with appendages inside at base; fruit large.
Ryparosa	Trees, shrubs or climbers; leaves without marginal glands, 3-veined, with pellucid dots; petioles swollen top and bottom; flowers unisexual; corolla with appendages inside at base; stamens usually fused; stigma broad and sessile.

Gesneriaceae

Contributed by Gemma Bramley

Field characters:

Herbs, lianas and shrubs; leaves opposite, sometimes anisophyllous; stipules absent; flowers tubular; ovary superior; fruit dehiscent or indehiscent; seeds many.

Description:

Habit herbs, lianas or shrubs, sometimes epiphytic.

Sap absent.

Stipules absent.

Leaves opposite, decussate, simple, margins entire to serrate, often anisophyllous (leaf pairs unequal), often hairy.

Inflorescences terminal or axillary.

Flowers zygomorphic, bisexual, tubular, 2-lipped; stamens 2 or 4.

Ovary superior.

Fruit dehiscent capsule or indehiscent berry; seeds many, usually <1 mm long.

Confused with:

Acanthaceae
Usually herbs, stems with swollen nodes, shrunken when dried; leaves usually with cystoliths; fruit usually with hooks inside; seeds 2–10.

Bignoniaceae
Mostly trees or lianas; fruit with many winged seeds.

Lamiaceae
Trees to herbs; fruit usually a drupe or mericarps; seeds 1–4(–7), seldom with wings.

Scrophulariaceae Usually herbs; leaves rarely densely hairy, sometimes drying blue-black; fruit with a persistent style; seeds many. More common in drier habitats.

Cyrtandra oblongifolia

Habit of *Aeschynanthus tricolor*

Rhynchoglossum sp.

Gesneriaceae cont.

Flower of *Aeschynanthus tricolor*

© RBG Kew

Genera:	
Aeschynanthus	Liana, mainly epiphytic; leaves opposite, rarely whorled, margins usually entire; 2-fertile stamens; fruit a linear loculicidally dehiscent capsule with seeds with appendages at both ends.
Agalmyla	Liana, usually a climber; leaves opposite, margins usually serrate; (2–)4 fertile stamens; fruit elongate, cylindrical, loculicidally and then septicidally dehiscent with arillate seeds.
Cyrtandra	Herb or shrub; leaves opposite; 2 fertile stamens; fruit an indehiscent berry.
Epithema	Herb; inflorescence densely compact, usually immersed in a bract; leaves opposite to whorled; 2 fertile stamens; fruit circumsessile dehiscent capsule.
Henckelia	Herb; leaves opposite or spirally arranged; 2 fertile stamens; fruit a straight capsule which opens on the upper side.
Hexatheca	Herb; leaves opposite or spirally arranged; 4 fertile stamens; fruit a cylindrical irregularly dehiscing capsule.
Monophyllaea	Herb; leaf 1, simple, large, ever-growing; 4 fertile stamens; fruit a capsule with 4 valves or 1 pore.
Paraboea	Herbs, only occurring on limestone; leaves opposite or spirally arranged; 2 fertile stamens; fruit a loculicidally dehiscent, spirally twisted, capsule or a straight capsule that opens on the upper side.
Rhynchotechum	Herbs or shrubs; leaves opposite; 4 fertile stamens; fruit an indehiscent berry.

Gnetaceae

Field characters:

Trees or climbers, dioecious, twigs with swollen nodes; leaves decussate, entire, penninerved; stipules absent; inflorescence a spike.

Description:

Habit trees or more often climbers; twigs with swollen nodes.

Sap absent.

Stipules absent.

Leaves decussate, margins entire, penninerved, often with translucent lines, dried leaves often black.

Inflorescence a spike; with small flowers arranged opposite each other.

Flowers small, dioecious.

Fruit drupe-like, orange, pink or yellow when mature.

Confused with:

Celastraceae
Swollen nodes absent; inflorescence never a spike; flowers with petals and sepals.

Guttiferae
White or yellow sap present; inflorescence never a spike; flowers with petals and sepals.

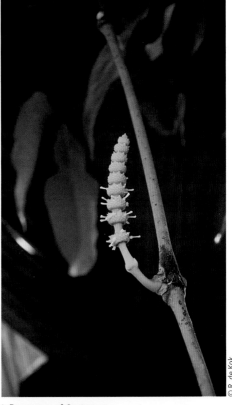

Inflorescence of *Gnetum* sp.

Habit of *Gnetum* sp.

Genus:

Gnetum	As for the family.

Icacinaceae

Field characters:

Trees, shrubs or lianas; leaves often spirally arranged, simple, margins entire; stipules absent; petals often free; ovary superior; fruit a drupe.

Description:

Habit trees, shrubs or lianas.

Sap absent.

Stipules absent.

Leaves opposite or spirally arranged; simple, margins entire to dentate, sometimes lobed, veins penni- or palminerved.

Flowers bi- or unisexual, regular, (3–)5(– 6)-merous; petals 4–6, free or connate at the base; stamens as many as petals and sepals; anthers opening by longitudinal slits.

Ovary superior, 1-locular, style 0–1.

Fruit a drupe, often laterally compressed; single-seeded.

Confused with:

Lauraceae

Flowers (2–)3-merous; anthers opening by valves; ovary with a single ovule.

Euphorbiaceae

Stipules present; flowers always unisexual, often 3-merous.

Fruit of *Gonocaryum littorale*

Fruit of *Iodes* sp.

© R. de Kok

© T. Utteridge

Fruit of *Phytocrene* sp.

© M. Coode

Flowers of *Gonocaryum littorale*

© T. Utteridge

Flowers of *Citronella suaveolens*

© T. Utteridge

Genera:	
Apodytes	Trees or shrubs; flowers bisexual; calyx cup-like; ovary and fruit with a lateral swelling or appendage; style excentric.
Citronella	Trees or shrubs; flowers bisexual; sepals free for $^3/_4$ of length and imbricate; petals free.
Gonocaryum	Trees or shrubs; petiole transversely ribbed; flowers unisexual; sepals free for $^3/_4$ of length and imbricate; petals free.
Gomphandra	Trees or shrubs; calyx cup-like; flowers unisexual, inflorescences in short cymes; drupe not laterally compressed and without a fleshy appendage; stamens hairy.
Iodes	Climber with tendrils; leaves opposite; fruit laterally compressed.
Miquelia	Climber; leaves spirally arranged, pinnately or palmately veined; flowers in heads or umbels; style absent.
Platea	Trees or shrubs; leaves with fine scales below, appearing shiny; flowers unisexual; sepals free for $^3/_4$ of length; petals connate; filaments free.
Phytocrene	Climber; leaves spirally arranged, palmately veined; flowers in heads or umbels; style short; fruit with stiff bristles or hairs.
Sarcostigma	Climber; leaves spirally arranged; flowers in elongated spikes.
Stemonurus	Trees or shrubs; flowers bisexual, sessile on a cymose inflorescence; calyx cup-like; stamens with long hairs.
Sleumeria	Climber; leaves spirally arranged, with yellow hairs, drying silvery grey, no tendrils; flowers on short racemes.

Lamiaceae

Contributed by Gemma Bramley

Field characters:

Leaves opposite; stipules absent; flowers usually 2-lipped, colourful; fruit drupes or mericarps, with an enlarged or colourful calyx.

Description:

Habit herbs, shrubs, trees or lianas.

Sap absent.

Stipules absent.

Leaves opposite, decussate, simple or palmately (rarely pinnately) compound; often with gland-tipped hairs or sessile glands.

Inflorescences determinate, terminal or axillary, cymose or thyrsoid.

Flowers bisexual, zygomorphic, rarely regular, usually tubular, 2-lipped; stamens 2 or 4, exceeding the corolla tube.

Ovary superior, style terminal or gynobasic.

Fruit with 2 carpels but often appearing 4-locular because a false division is formed by the ingrowth of the carpel wall; 1 ovule per locus; seeds (2–)4 nutlets or in a drupe containing 1–4(–7) stones.

Confused with:

Rubiaceae

Fused interpetiolar stipules present; flowers regular; ovary usually inferior.

Solanaceae

Leaves usually alternate; stamens 5, opening often with pores; fruit a berry, seeds many.

Apocynaceae

White sap present; flowers with twisted petals, androecium and gynoecium sometimes fused; fruit paired, seed usually with tuft of hairs.

Rutaceae

Leaves often alternate, often compound; flowers regular, corolla and calyx lobes free.

Fruit of *Teijsmanniodendron coriaceum*

Flower of *Vitex vestita*

Clerodendrum sp.

Premna serratifolia

Glands at base of *Gmelina aborata* leaf

Callicarpa longifolia

Swollen nodes of *Teijsmanniodendron sinclairii*

Genera:	
Clerodendrum	Small trees or shrubs; leaves simple; corolla regular to zygomorphic, 5-lobed; fruiting calyx usually recurved.
Callicarpa	Small trees or shrubs; leaves simple; inflorescence axillary, often with stellate hairs; corolla regular, 4(–5)-lobed.
Congea	Liana; leaves simple; inflorescence with 3–4 involucral bracts; corolla zygomorphic.
Gmelina	Trees, shrubs or lianas; leaves simple, with 2 large basal glands; large showy corolla.
Gomphostemma	Herbs; leaves simple, often with stellate hairs; corolla zygomorphic, showy.
Hyptis	Weedy herbs; leaves simple; inflorescence congested cymes or involucrate capitula; corolla slightly zygomorphic; stamens declinate.
Peronema	Trees; leaves pinnate, often with winged leaf rachis; 2 fertile stamens.
Petraeovitex	Lianas; leaves palmate; flower regular; fruiting calyx with 5 greatly enlarged lobes (wings).
Plectranthus	Herbs; leaves simple; corolla zygomorphic, sigmoid-shaped; stamens declinate.
Premna	Trees, shrubs or lianas; leaves simple; inflorescences terminal; corolla regular to zygomorphic, 4-lobed; fruit with 1–4 seeds.
Sphenodesme	Lianas; leaves simple; inflorescences with 6 involucral bracts; corolla regular.
Tectona	Big trees; leaves simple, large, with stellate hairs; large inflated papery calyx when in fruit.
Teijsmanniodendron	Trees; leaves simple or palmately compound, petioles with swollen nodes; corolla regular to zygomorphic, 5-lobed; fruit with 1 seed.
Vitex	Trees or shrubs; leaves palmate; corolla regular to zygomorphic, 5-lobed; fruit with 1–4 seeds.

Lauraceae

Field characters:

Leaves alternate or spirally arranged, simple, sometimes aromatic when crushed, margins entire; stipules absent; flowers 3-merous; anthers opening by valves; fruit a single-seeded berry.

Description:

Habit trees or shrubs or twining parasite.

Sap absent.

Stipules absent.

Leaves alternate or spirally arranged, or rarely sub-opposite or pseudo-whorled, simple, margins entire, venation orthogonal-reticulate, glaucous below, sometimes aromatic when crushed.

Inflorescences axillary, occasionally pseudo-terminal; usually branched, enveloped in bracts or surrounded with an involucre of bracts.

Flowers bisexual or unisexual, (2–)3-merous, very small, yellowish, greenish or white; tepals in 2 whorls; stamens usually in 4 whorls with inner whorl sterile; anthers 2- or 4-locular, opening by valves.

Ovary usually superior, with a single ovule.

Fruit a berry, receptacle or pedicel often enlarged and colourful (red, orange or yellow), sometimes entirely enclosing the fruit; single-seeded.

Confused with:

Icacinaceae
Climbers or shrubs; flowers 5-merous, anthers not opening by valves.

Monimiaceae
Leaves (sub-)opposite, margins dentate; fruit many, sitting on a torus.

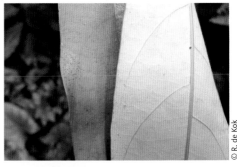
© R. de Kok

Glaucous lower surface

© R. de Kok

Fruit of *Dehaasia* sp.

© T. Utteridge

Habit of *Litsea* sp.

© RBG Kew

Flower of *Litsea* sp. with anther flaps

© T. Utteridge

Flower of *Endiandra macrophylla*

Genera:	
Actinodaphne	Trees or shrubs with a *Terminalia*-type branching pattern; leaves often in whorls rarely spirally arranged, crowded at apex, bud scales not conspicuous; flower surrounded by large involucre bracts; perianth lobes 6, unequal; anthers 4-celled; fruit sitting on a cup-shaped perianth tube.
Alseodaphne	Trees and shrubs; leaves spirally arranged, rarely alternate; perianth lobes 6 in 2 whorls, outer lobes usually shorter than inner; stamens 9 in 1 or 2 rows; anthers 4-celled; fruit stalks swollen and fleshy.
Beilschmiedia	Trees with *Terminalia*-type branching pattern; leaves spirally arranged or alternate; perianth lobes 6 in 2 whorls; lobes sub-equal; stamens 9 in 3 rows, anthers 2-celled; fruit stalks not swollen or colourful.
Cassytha	Parasitic climber; stem thin, yellowish; leaves absent; ovary inferior.
Cinnamomum	Trees and shrubs; leaves mostly opposite or sub-opposite, 3-veined from the base; perianth lobes 6 in 2 whorls; lobes sub-equal; stamens 9 in 3 rows, anthers 4-celled; fruit sitting on a cup-shaped perianth tube.
Cryptocarya	Trees or shrubs; leaves alternate; perianth lobes 6 in 2 whorls; lobes sub-equal; stamens usually 9 in 3 rows, anthers 2-celled; mature fruit entirely enclosed by the receptacle and appearing inferior.
Dehaasia	Trees or shrubs; leaves spirally arranged, usually drying blackish; perianth lobes 6 in 2 whorls; outer lobes much smaller; stamens 9 in 3 rows, anthers 2-celled; fruit stalks swollen, fleshy and colourful.
Endiandra	Trees or shrubs; leaves alternate, with distinct areolate reticulations; perianth lobes 6 in 2 whorls; inner lobes smaller; stamens 3 in 1 row, anthers 2-celled; fruit without a perianth.
Eusideroxylon	Trees; leaves alternate; perianth lobes 6, sub-equal; stamens 3 in 1 row; anthers 4-celled; fruit stalks not swollen.
Lindera	Small trees or shrubs; leaves alternate to opposite, usually drying blackish; flower surrounded by large involucre bracts; perianth 6-lobed; stamens 9 or 12, in 3 or 4 rows, anthers 2-celled; fruit sitting on a slightly enlarged perianth tube.
Litsea	Trees or shrubs; leaves spirally arranged to opposite, bud scales not conspicuous; flower surrounded by large involucre bracts; perianth lobes (0–)6; stamens 9 or 12, in 3 or 4 rows; anthers 4-celled; fruit sitting on a enlarged perianth tube, sometimes appearing cupular.
Neolitsea	Trees or shrubs; leaves spirally arranged to opposite; flower surrounded by large involucre bracts; perianth lobes 4; stamens 6 in 3 rows; anthers 4-celled; fruit sitting on a slightly enlarged perianth tube.
Nothophoebe	Trees or shrubs; leaves spirally arranged to opposite; perianth lobes 6 in 2 whorls, outer lobes smaller than inner; stamens 9 in 1 or 2 rows, anthers 4-celled; fruit sitting on a slightly enlarged perianth tube.
Persea	Trees or shrubs; leaves spirally arranged; perianth lobes 6 in 2 whorls; lobes equal, not woody, spreading or reflexed in fruit; stamens 9 in 1 or 2 rows, anthers 4-celled; fruit sitting on hardly enlarged spreading perianth lobes.
Phoebe	Trees with *Terminalia*-type branching pattern; leaves alternate or spirally arranged; perianth lobes 6 in 2 whorls; fruit lobes equal, woody, cup-shaped; stamens 9 in 1 or 2 rows, anthers 4-celled; fruit sitting on a cup-shaped perianth tube.

Lecythidaceae

Field characters:

Trees or shrubs; leaves spirally arranged, simple, often obovate, margins minutely and regularly toothed, penninerved; stamens numerous; ovary (semi-)inferior.

Description:

Habit trees or shrubs.

Sap absent.

Stipules small or absent.

Leaves spirally arranged or whorls, often crowded at the tips of branches; simple, often obovate; margins minutely and regularly toothed to serrate, penninerved.

Flowers bisexual, large, petals often clawed, stamens numerous; on long pendulous inflorescences.

Ovary (semi-)inferior, 2–6-locular; style long, fine often persistent on fruit.

Fruit a capsule, often ridged, with fibrous mesocarp.

Confused with:

Combretaceae

Trees or climbers, leaves spirally arranged or opposite, margins entire, with scales, often with pellucid dots; ovary 1-locular.

Myrtaceae

Leaves mostly opposite, margins entire, with pellucid dots, often intramarginal vein present.

© RBG Kew

Barringtonia macrostachya

© RBG Kew

Barringtonia edulis

Genera:

Barringtonia	Leaves in whorls or clustered at the end of the twigs, base not decurrent on the petiole; petiole swollen at base; inflorescence pendulous and generally long (>20 cm long).
Planchonia	Leaves spirally arranged, well placed along the twigs, base decurrent on the petiole; petiole not swollen at base; inflorescence mostly erect and generally short (<15 cm long).

Leeaceae

Field characters:

Leaves spirally arranged, margins dentate, base of the petiole expands to form a clasping stipular structure; inflorescence situated opposite the leaf; flowers regular, stamens forming a tube that is joined to the corolla.

Description:

Habit trees, shrubs or lianas, sometimes stout herbs; sometimes with spines, tendrils absent.
Sap absent.
Stipules base of the petiole expands to form a clasping stipular structure.
Leaves spirally arranged; 1–4 pinnate or simple, usually imparipinnate; margins dentate, rachis sometimes winged, pinnately veined.
Flowers bisexual, regular, 4–5-merous; petals jointed at base; stamens form a tube that is joined to the corolla.
Ovary superior, 4–8-locular, 1 ovule per cell.
Fruit a berry, apex depressed.

Confused with:

Vitaceae
Climbers with tendrils; stamens free.

Leea indica

© R. de Kok

Fruit of *Leea* sp.

© J. Gregson

Genus:

Leea	As for the family.

Leguminosae–Caesalpinioideae

Contributed by Ruth Clark

Field characters:

Trees, shrubs or lianas, sometimes with thorns; leaves usually spirally arranged, usually pinnate or bipinnate; flowers large, petals the showiest part of the flower, the median petal is overlapped by the other petals; ovary superior; fruit usually a legume.

Description:

Habit trees, shrubs or lianas, sometimes with thorns.

Sap absent.

Stipules usually present, but often falling early.

Leaves usually spirally arranged, usually pinnate or bipinnate, more rarely unifoliolate; petiole usually swollen at base; pulvini present for 'sleep movements'.

Flowers bisexual, zygomorphic, but often appearing regular, usually 5-merous, often large; petals the most showy part of the flower; median petal is overlapped by the 4 others; sepals free; stamens free, up to 10, often with hairs at base.

Ovary superior, 1-locular.

Fruit usually a legume, sometimes drupaceous; dehiscent or indehiscent, sometimes winged

Confused with:

Connaraceae
Stipules absent; sepals free; fruit often consist of c. 5 separate carpels.

Leguminosae–Papilionoideae
Median petal overlaps the 2 adjoining petals.

Flower of *Bauhinia kockiana* var. *kockiana*

Phanera (*Bauhinia*) *integrifolia* subsp. *cumingiana*

Major genera:

Afzelia	Trees, unarmed; leaves paripinnate; flowers with only 1 developed petal; fertile stamens 5–7; fruit dehiscent; seed arillate.
Caesalpinia s.l.	Climbers, shrubs or trees, usually armed; leaves bipinnate, usually spiny; flowers with 5 petals; fertile stamens 10; fruit dehiscent or indehiscent, sometimes winged along 1 suture; seed not arillate.
Cassia s.l.	Trees or large shrubs, unarmed; leaves paripinnate; flowers with 5 petals; fertile stamens 10 of which 3 are twice as long as the others; fruit indehiscent; seed not arillate.
Crudia	Trees or shrubs, rarely scramblers, unarmed; leaves imparipinnate; petals absent or very small, calyx lobes 4; fertile stamens 10; fruit a dehiscent wingless legume; seed not arillate.
Cynometra	Trees or shrubs, unarmed; leaves paripinnate; petals 5; fertile stamens 10, equal in length; fruit indehiscent; seed not arillate.
Dialium	Trees, unarmed; leaves imparipinnate; petals absent or small; calyx lobes 3 or 5; fertile stamens 2 or 6; fruit indehiscent, drupaceous, wingless; seed surrounded by a sweet pulp.
Intsia	Trees; leaves paripinnate, unarmed; petals 1; fertile stamens 3; fruit dehiscent with leathery valves; seed not arillate.
Koompassia	Trees with smooth bole, unarmed; leaves imparipinnate; petals 5; fertile stamens 5; fruit indehiscent or slightly dehiscent, surrounded by a papery wing; seed not arillate.
Phanera (*Bauhinia*)	Climbers with tendrils or shrubs or trees; leaves unifoliolate, leaflet either bilobed or deeply divided; petals (1 –)5(– 6); fertile stamens 0–10; fruit (in)dehiscent with woody or leathery valves; seed not arillate.
Peltophorum	Climbers or shrubs, unarmed; leaves bipinnate; petals 5, yellow; fertile stamens 10; fruit indehiscent, a samara; seed not arillate.
Saraca	Trees or shrubs, unarmed; leaves paripinnate; petals absent; fertile stamens (3–)4–8(–10); fruit dehiscent with twisting valves; seed not arillate.
Sindora	Trees, unarmed; leaves paripinnate; calyx lobes acute; petals 1; fertile stamens 10, uppermost 1 free and reduced; fruit dehiscent, usually spiny; seed arillate.
Sympetalandra	Trees, unarmed; leaves pinnate or bipinnate, paripinnate; petals 5; stamens 10, alternately long and short; fruit dehiscent with woody valves; seed not arillate.

Leguminosae–Mimosoideae

Contributed by Ruth Clark

Field characters:

Trees, shrubs or lianas, sometimes with thorns or prickles; leaves alternate, usually bipinnate; flowers small, in heads or spikes, these often aggregated into a compound inflorescences, stamens the showiest part of the flower; ovary superior; fruit usually a pod, sometimes a craspedium.

Description:

Habit trees, shrubs or lianas, sometimes with thorns or prickles.

Sap absent.

Stipules rarely absent, but often falling early.

Leaves alternate, bipinnate or appearing simple (phyllode-leaf common in *Acacia*); usually with extra-floral nectaries on the petiole, rachis or pinnae; pulvini present for 'sleep movements'; sometimes sensitive to touch.

Flowers bisexual, regular, usually 5-merous, small; sepals and petals valvate; stamens the showiest part of the flower, 10–many, free or united into a tube.

Ovary superior, 1-locular, placentation parietal.

Fruit usually a legume, sometimes a craspedium.

Confused with:

Leguminosae–Caesalpinioideae

Flowers usually large; petals forming the showiest part of flower.

© S. Marsh

Mimosa pudica

Falcataria moluccana

Major genera:

Acacia	Trees, shrubs or lianas, sometimes armed; leaves apparently simple, usually with prominent parallel veins, or rarely bipinnate with opposite leaflets, not sensitive to touch; rachis or petiole of leaves usually with nectaries.
Albizia	Trees, shrubs or lianas, unarmed apart from lianas; leaves bipinnate, leaflets opposite, not sensitive to touch; rachis of leaves with nectaries.
Adenanthera	Trees or shrubs, unarmed; leaves bipinnate, leaflets alternate, not sensitive to touch; rachis of leaves without nectaries.
Archidendron	Trees or shrubs, unarmed; leaves bipinnate, with usually opposite leaflets, not sensitive to touch; rachis of leaves with nectaries.
Entada	Lianas or shrubs with tendrils, unarmed; leaves bipinnate, leaflets mostly opposite, not sensitive to touch; rachis of leaves without nectaries.
Mimosa	Trees, shrubs or climbers, sometimes armed; leaves bipinnate, leaflets opposite, sometimes sensitive to touch; rachis of leaves without nectaries.
Neptunia	Herbs, unarmed; bipinnate leaves, leaflets opposite, sensitive to touch; rachis of leaves with or without nectaries.
Parkia	Trees, unarmed; leaves bipinnate, leaflets opposite, not sensitive to touch; rachis of leaves usually with nectaries; inflorescences in pendulous clavate heads.
Serianthes	Trees or shrubs, unarmed; leaves bipinnate, leaflets alternate, not sensitive to touch; rachis of leaves usually with nectaries; inflorescences in spikes.

Leguminosae–Papilionoideae

Contributed by Ruth Clark

Field characters:

Trees, shrubs, lianas or herbs; leaves alternate, unifoliolate to pinnate or pinnately trifoliolate; petals the most showy part of the flower, median petal overlaps the 2 adjoining petals; fruit usually a legume or a loment.

Description:

Habit trees, shrubs, lianas or herbs.

Sap rarely present.

Stipules usually present, but often falling early.

Leaves alternate, pinnate to pinnately trifoliate or unifoliolate; pulvini present on petiole for 'sleep movements'; leaflet margins usually entire.

Flowers bisexual, zygomorphic, usually 5-merous; sepals united at base into a calyx tube; petals the showiest part of flower, median petal overlaps the 2 adjoining petals; stamens 10–many, free or united into a sheath.

Ovary superior, 1-locular.

Fruit a legume.

Confused with:

Connaraceae
Stipules absent; sepals free; fruit often consist of c. 5 separate carpels.

Leguminosae–Caesalpinioideae
Median petal overlapped by the other petals; stamens free, often hairy.

Polygalaceae
Leaves simple; stipules absent; ovary usually 2-locular.

© M. Coode

© J. Gregson

Fruit of *Ormosia* sp.

Habit of *Sesbania grandiflora*

Flowers of *Mucuna* cf. *hainanensis* subsp. *multilamellata*

© J. Gregson

Major genera:

Crotalaria	Trees, or more commonly shrubs or herbs; leaves simple to 1–3(–7) palmate; stipules present or absent; fruit dehiscent, inflated.
Dalbergia	Trees, shrubs or lianas, sometimes spiny, red sap sometimes present; leaves unequally pinnate with alternate leaflets; fruit indehiscent, winged.
Desmodium	Herbs; leaves pinnate, 1–3(–5) leaflets; fruit an (in)dehiscent loment, straight along one margin and sinuous along the other.
Eriosema	Shrubs or herbs; leaves pinnately trifoliolate, with orange glands on lower surface; legume dehiscent, with many long hairs.
Indigofera	Shrubs or herbs; leaves pinnate, leaflets 1–many, with T-hairs especially on the lower surface; pods cylindrical.
Fordia	Small trees or shrubs; leaves pinnate; axillary buds usually a little way above the leaf axial; flowers along the main stem; legumes with 2 seeds.
Millettia	Trees or climbers, red sap sometimes present; leaves simple or paripinnate; leaflets opposite; legumes woody, dehiscent; seeds 1–4, massive.
Mucuna	Lianas; leaves trifoliolate; legumes usually dehiscent, with irritant hairs.
Spatholobus	Lianas; leaves trifoliolate; inflorescence large, paniculate; fruit indehiscent, like a samara.
Tephrosia	Shrubs or herbs; leaves pinnate, leaflets 1–many, with close secondary venation; fruit usually a dehiscent, hairy legume.
Rhynchosia	Shrubs or herbs, sometimes climbing; leaves pinnately, trifoliolate, with orange glands on lower surface; legume dehiscent, with short hairs.

Loganiaceae

Field characters:

Leaves opposite, sometimes in whorls, simple; stipules absent, but interpetiolar ridge present; inflorescence terminal; flowers regular, petals fused, stamens 4–5; ovary superior.

Description:

Habit trees, shrubs, lianas, rarely herbs.
Sap absent.
Stipules absent, but with an interpetiolar ridge present that sometimes resembles an interpetiolar stipule, or an ochrea or its scar.
Leaves opposite, sometimes in whorls, simple, margins usually entire.
Flowers bisexual, regular, (4–)5-merous; petals fused; stamens 4–5, alternating with corolla lobes.
Ovary superior, (1–)2(–4)-locular; usually 1 style, stigma club-shaped to deeply bifid.
Fruit a berry, capsule or drupe; seeds 1–many per fruit.

Confused with:

Apocynaceae
White sap present; fruit in pairs.
Rubiaceae
Interpetiolar stipules present, ovary usually inferior.

Fruit of *Fagraea cuspidata*

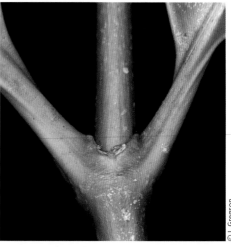

Fagraea cuspidata with basal excavations

Habit of *Fagraea* sp.

Genera:

Fagraea	Trees, shrubs or lianas; leaves pinnately nerved; petiole with excavations at base; fruit a berry.
Gelsemium	Lianas; leaves pinnately nerved; petiole without excavations at base; style twice forked; fruit a capsule.
Mitreola	Herbs; leaves pinnately nerved, with interpetiolar stipules or ridges; flowers 5-merous; fruit a capsule.
Norrisia	Trees; leaves pinnately nerved; petiole without excavations at base; style slightly 2 lobed or club-shaped; fruit a capsule.
Strychnos	Small trees, shrubs or lianas usually with tendrils in leaf axis; leaves with 3–5 main veins from base; fruit a berry.

Loranthaceae

Field characters:

Hemi-parasitic shrubs growing on tree branches; leaves simple, margins entire; stipules absent; calyx reduced; corolla well developed; stamens opposite the petals; ovary inferior; fruit single-seeded berries.

Description:

Habit hemi-parasitic shrubs, attached to the branches of trees with a haustoria (usually a swollen structure clasping tree branches or attached to them).

Sap absent.

Stipules absent.

Leaves usually opposite, simple, margins entire, venation usually indistinct; leaves often turn black when drying.

Flowers usually bisexual; calyx reduced; corolla usually fused and always well developed; stamens as many as, and opposite, the petals.

Ovary inferior.

Fruit a single-seeded berry; seed surrounded by a slimy layer.

Confused with:

Olacaceae
Leaves rarely opposite, surface finely tuberculate; corolla usually small; ovary superior.

Proteaceae
Trees; leaves often serrate; no separate calyx and corolla; ovary superior.

Santalaceae
Small trees or shrubs, not epiphytic; seeds not surrounded by slimy layer.

Viscaceae
Flowers unisexual, with tiny petals.

Fruit of *Lampas elmerii*

Haustoria of *Macrosolen cochinchinensis*

Haustoria of *Lampas elmerii*

© T. Utteridge

© R. de Kok

Loranthaceae cont.

© J. Gregson

Flowers of *Lepeostegeres centiflorus*

Genus:	
Amyema	Leaves opposite to verticillate, 3-veined from base; inflorescence a head surrounded by bracts; petals free.
Dendrophthoe	Leaves spirally arranged; inflorescence not a head surrounded by bracts; petals fused.
Helixanthera	Leaves spirally arranged or in whorls; inflorescence resembling a spike, not a head surrounded by bracts; flowers not in triads; petals free.
Lampas	Leaves in whorls; inflorescence a head surrounded by bracts; petals fused.
Lepeostegeres	Leaves mostly opposite, inflorescence a head surrounded by red bracts; petals fused.
Lepidaria	Leaves opposite; inflorescence a head surrounded by red bracts; petals free.
Macrosolen	Leaves opposite; inflorescence not a head or surrounded by bracts; petals fused.
Scurulla	Leaves opposite, young parts rusty hairy, inflorescence not a head or not surrounded by bracts; flowers not in triads; petals fused; fruit club-shaped.

Magnoliaceae

Field characters:

Trees; leaves spirally arranged, simple, margins entire, penninerved; stipules present, leaving a circular scar; flowers single, tepals free, stamens usually numerous, carpels numerous.

Description:

Habit trees.

Sap absent.

Stipules present, leaving a circular scar.

Leaves spirally arranged, simple, margins entire, penninerved, sometimes glaucous.

Flowers bisexual, single, usually large, actinomorphic, aromatic; tepals free, several series in a spiral, imbricate; stamens usually many, free.

Ovary superior, carpels numerous, free in upper part.

Fruit usually woody, dehiscent.

Confused with:

Annonaceae
Tree, shrubs or climbers; twigs in cross-section with rays in the wood; stipules absent.

Lauraceae
Stipules absent; flowers small, anthers and carpels not free.

Winteraceae
Stipules absent; flowers unisexual, carpels free.

© R. de Kok

Circular scar of *Magnolia* sp.

© R. de Kok

Fruit of *Magnolia* sp.

Genus:

Magnolia	As for the family.

Malvaceae

Field characters:

Trees, shrubs or herbs; leaves spirally arranged, simple, often with stellate hairs; flowers regular, calyx connate, often with an epicalyx; petals free; stamens numerous, filament fused in a tube and fused at the base with petals; ovary superior.

Description:

Habit trees, shrubs and herbs; bark usually very fibrous.

Sap absent, rarely present, then white or yellow.

Stipules present.

Leaves alternate, simple, margins entire to deeply lobed; usually palmately veined or 3-nerved from the base; stellate hairs often mixed with simple hairs, rarely with peltate scales.

Flowers bisexual, 5-merous, regular, showy; sepals usually connate, but often nearly free, often with an epicalyx; petals are convolute or imbricate; stamens many, united for most of their length into a cylinder around the style, fused at base with the corolla.

Ovary superior, with 1–many locules.

Fruit capsules or a schizocarp that breaks into 5 single- (rarely 2-)seeded mericarps, often equipped with awns; seed often hairy.

Confused with:

Durionaceae (Bombacaceae)
Trees, leaves often penninerved, often with dense scales on the lower surface; epicalyx absent; seeds often glabrous.

Sparrmanniaceae
Leaves 3-veined from the base, often with domatia; stamens free or rarely united in bundle(s), then not fused to the corolla.

Sterculiaceae
Flowers often with a single whorl of perianth; stamens free or fused in a bundle, not fused to the corolla; fruit smooth, carpels free from each other.

© RBG Kew

Calyx and epicalyx of *Hibiscus surratensis*

© RBG Kew

Habit of *Hibiscus surratensis*

© R. de Kok

Flower of *Hibiscus* sp. with staminal tube

Genera:	
Abutilon	Herbs and shrubs; flowers >2.5 cm diameter; epicalyx absent; fruit a schizocarp with 2 seeds per mericarp.
Abelmoschus	Herbs or small shrubs; flowers >2.5 cm diameter; calyx not lobed, epicalyx present; fruit a capsule.
Hibiscus	Shrubs and trees; leaves often with scales or with glands on midrib; flowers usually >2.5 cm diameter; calyx 5-lobed, epicalyx present; fruit a capsule.
Gossypium	Shrub with a 3-lobed epicalyx; flowers >2.5 cm diameter; fruit a capsule with seed covered in long dense hairs (cotton).
Sida	Herbs and shrubs; flowers <2.5 cm diameter; epicalyx absent; fruit a schizocarp, 1 seed per mericarp.
Thespesia	Trees or shrubs, white or yellow sap sometimes present; leaves with scales sometimes present, petiole swollen apically; flowers >2.5 cm diameter; epicalyx present; fruit a capsule; seed glabrous or with short hairs.
Urena	Herbs and shrubs; flowers <3 cm diameter, epicalyx present; fruit a schizocarp, 1 seed per mericarp.

Marantaceae

Field characters:

Rhizomatous herbs, non-aromatic; leaves 2-ranked, simple, petiolate, sheathing at base and swollen apically; apex of leaf often offset from the midrib; blade often corrugated when dried.

Description:

Habit herbs, non-aromatic, glabrous; rhizomatous.

Sap absent.

Stipules absent.

Leaves 2-ranked, simple, margins entire; petiole sheathing at base and swollen apically; apex of leaf often offset from the midrib; blade often corrugated when dried.

Flowers bisexual or unisexual, when unisexual then female flower below male on spadix; zygomorphic, perianth in 3 rows of 3 lobes, unequal, more-or-less free; stamens petaloid, the inner 1 with a 1-locular anther.

Ovary inferior, 3-locular

Fruit a capsule; seeds usually few, often arillate.

Confused with:

Zingiberaceae
Aromatic; leaves with an open sheath, ligule present.

Orchidaceae
Stamens and style forming a column; roots covered in papery or velvet sheath.

© J. Gregson

Flower of *Phrynium* cf. *pubinerve*

© R. de Kok

Fruit of *Phrynium capitatum*

Habit of *Phrynium* cf. *pubinerve*

© J. Gregson

Genera:

Donax	Leaves clustered on top of a cane-like stem; inflorescence with small bracteoles; corolla tube the same length as the lobes or shorter; fruit indehiscent.
Phacelophrynium	Leaves in rosettes; inflorescence without bracteoles; sepals the same length as the corolla tube; fruit dehiscent.
Phrynium	Leaves in rosettes; inflorescence without bracteoles; sepals at least half the length of the corolla tube, usually longer; fruit dehiscent.
Schumannianthus	Leaves clustered on top of a cane-like stem; inflorescence with small bracteoles; corolla tube half the length of the lobes or less; fruit dehiscent.
Stachyphrynium	Leaves in rosettes; inflorescence without bracteoles; sepals one third the length of the corolla tube or shorter; fruit dehiscent.

Melastomataceae

Contributed by Eve Lucas

Field characters:

Plants usually woody; leaves simple, opposite, often 3–5(–7) parallel-veins from the base; stipules absent; stamens usually 8, sickle-shaped, isomerous, with appendages; ovary inferior or semi-inferior.

Description:

Habit trees, epiphytes and shrubs; rarely herbs; twigs without thickened nodes.

Sap absent.

Stipules absent, rarely pseudo-stipules present.

Leaves opposite, rarely in whorls, simple, margins usually entire, sometimes crenate, usually without pellucid dots, 3–5(–7) parallel veins from base.

Flowers usually hermaphroditic, hypanthium often bearing stellate hairs; petals often showy and colourful; stamens usually 8, rarely 3, sickle-shaped, isomerous, with appendages; anthers often opening with pores.

Ovary (semi-)inferior, usually 4–5-celled with numerous axile ovules.

Fruit berries or dry capsules.

Confused with:

Myrtaceae

Leaves with pellucid dots; inflorescences not generally cymose; flowers never blue or purple; stamens (and stamen scars) numerous, without appendages.

Crypteroniaceae

Twigs thickened at the nodes; petals usually minute or absent; anthers without appendages.

Fruit with star-shaped ridges of *Memecylon* sp.

© J. Gregson

Flower of *Melastoma malabatricum*, the stamens with appendages

© R. de Kok

© J. Gregson

Habit of *Pternandra* sp.

Major genera:

Anerincleistus	Shrubs; leaves anisophyllous; stamens 8, equal and isomorphous; capsule loculicidally dehiscing at the top with 4 valves.
Astronia	Shrubs or trees; leaves with pellucid dots, usually with scales, stellate hairs or cystoliths; stamens 10, equal and isomorphous; capsule dehiscing irregularly revealing vascular bundles as a stellate frame.
Blastus	Herbs or shrubs; leaves with glands on the lamina, sometimes anisophyllous; stamens 4, equal; capsule dehiscing at the top with 4 valves.
Dalenia	Climber; stem with a raised ridge between the petioles; leaves 7-nerved; stamens 8, unequal, 4 large and linear, 4 small and S-shaped; fruit an ovoid berry.
Diplectria	Climber; stem with a raised ridge between the petioles; leaves 3- or 5-nerved; stamens 8, mostly very unequal, 4 large with 1 or 2 dorsal appendages, 4 small, with or without dorsal appendages but with 2 linear appendages; fruit a sub-globose berry.
Dissochaeta	Climber; stem with a raised ridge between the petioles, sometimes with swollen nodes, cinereous stellate hairs present and sometimes with scales; inflorescences long and open panicles; stamens 4, equal; fruit an elliptic berry.
Medinilla	Climbers or shrubs, often epiphytes; leaves often fleshy; inflorescences mostly terminal or axillary panicles or cymes; stamens 8–12, equal or unequal; fruit a sub-globose, thin-walled berry.
Melastoma	Shrubs or small trees, usually with dendroid hairs; stamens 10(–14); anthers isomorphous, large anthers purple, with long, arcuate appendices on the connective, small anthers yellow, ventrally bilobate, connective not produced; fruit a coriaceous or fleshy berry, irregularly dehiscent or indehiscent.
Memecylon	Small trees to shrubs; leaves usually glabrous, midnerve and marginal veins usually prominent, secondary nerves strong to absent; inflorescences axillary umbels or cymes; stamens 8, equal; flower disc often bearing 8 star-shaped ridges and 8 staminal scars; fruit 1-locular; 1–2-seeded.
Pachycentria	Epiphyte with swollen stems, glabrous; leaves often fleshy; inflorescences terminal or axillary panicles; stamens 8–10, equal; fruit urn-shaped berries.
Pternandra	Small trees with swollen nodes, glabrous; inflorescences axillary panicles or fascicles; stamens 8, equal; fruit a sub-globose berry with simple or ornate armature.
Sonerila	Herbs; leaves spirally arranged, sometimes anisophyllous leaves, mostly sparsely hirsute; flowers 3-merous, solitary, in triads, racemes or scorpioid spikes; stamens 3–6, equal or sub-equal; capsule tuberculate or glabrous, widened and flattened at the top and rimmed by 3 scales, dehiscing with 3 valves.

Meliaceae

Field characters:

Trees, rarely shrubs; leaves spirally arranged, usually pinnately compound, usually imparipinnate; stipules absent; stamens united and forming a tube; ovary superior.

Description:

Habit trees, rarely big shrubs or scrambling small shrubs.

Sap absent.

Stipules absent, rarely with pseudostipules.

Leaves spirally arranged, usually pinnate to bipinnate, rarely simple, usually imparipinnate; leaflets alternate or sub-opposite, margins usually entire; sometimes ending in a free rachis tip; hairs simple to bifid or stellate, peltate scales sometimes present.

Flowers bisexual or unisexual, small; sepals usually 3–5; petals usually 3–7; disk present; stamens usually fused in a tube.

Ovary superior.

Fruit a capsule, berry or drupe, often globose; seeds dry with wings or fleshy and arillate.

Confused with:

Anacardiaceae
White sap present, turning black on exposure; stamens free.

Burseraceae
Resiniferous; leaves imparipinnate, often with pseudo-stipules; leaflets opposite and with distinct petiolules; stamens free, 2-ovules per cell.

Rutaceae
Leaves with pellucid dots; stamens free.

Sapindaceae
Leaves often with a free rachis tip; stamens free, hairy; fruit often 3-lobed or hairy.

© R. de Kok

Free rachis tip of *Chisocheton lasiocarpus*

© J. Gregson

Flower of *Dysoxylum cauliflorum*

© R. de Kok

© J. Gregson

Fruit of *Heynea trijuga*

Fruit of *Chisocheton* sp.

Inflorescence of *Aglaia oligophylla*

© J. Gregson

Genera:

Aglaia	Leaves imparipinnate, rarely simple, drying pale green, usually with scales or stellate hairs; fruit a berry.
Aphanamyxis	Leaves imparipinnate, with scales or simple hairs; fruit a capsule.
Azadirachta	Leaves paripinnate, hairs simple; fruit a drupe.
Chisocheton	Leaves imparipinnate, rarely paripinnate or simple, usually the rachis with a free tip, hairs simple; fruit a capsule.
Dysoxylum	Leaves imparipinnate, rarely simple, usually the rachis without a free tip, hairs simple; fruit a capsule.
Heynea	Leaves imparipinnate without a free end tip, hairs simple; fruit a capsule.
Lansium	Leaves paripinnate; hairs simple; fruit a berry, on branches and trunks.
Sandoricum	Leaves trifoliolate, hairs simple; fruit a drupe.
Toona	Leaves pinnate, usually paripinnate, hairs simple, stamens free; fruit a woody capsule with winged seeds.
Vavaea	Trees with *Terminalia*-type branching and simple leaves, hairs simple; fruit a berry.
Walsura	Leaves imparipinnate, rachis swollen at insertion of leaflets, hairs simple or bifid; fruit a 1–2(– 4)-seeded berry.
Xylocarpus	Mangrove trees; leaves paripinnate, without a free end tip, hairs simple, calyx valvate; fruit a large sub-spherical capsule.

Menispermaceae

Field characters:

Lianas, rarely shrubs, stems in cross-section with radial cavities; leaves spirally arranged, simple; petiole thickened apically; stipules absent; fruit a drupe, seeds often horseshoe-shaped.

Description:

Habit lianas or shrubs; lianas often in cross-section full of cavities, tendrils absent.

Sap sometimes with white sap.

Stipules absent.

Leaves spirally arranged, simple, margins usually entire, palminerved or penninerved; petiole thickened apically.

Inflorescences often extra-axillary or cauliflorous.

Flowers dioecious, often 3-merous, small, parts free and rarely brightly coloured.

Ovary superior, usually 3–many carpels per flower.

Fruit a drupe; seeds often horseshoe-shaped, endocarp variously ornamented.

Confused with:

Aristolochiaceae
Leaves withering away on the stem and leaving no scar; flowers usually tubular; fruit dehiscent.

Cucurbitaceae
Tendrils present; petiole apex not thickened; ovary inferior.

Passifloraceae
Tendrils present; glands on leaf and petiole; petiole apex not thickened; seeds numerous.

Dioscoreaceae
Petiole apex not thickened; fruit dry, usually winged.

Cross-section of stem with cavities of *Tinomiscium petiolare*

© R. de Kok

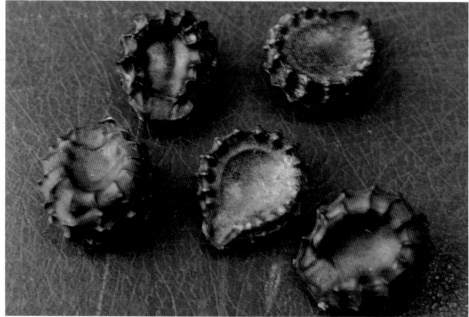

Dried fruit with horseshoe-shaped seeds of *Stephania corymbosa*

© R. de Kok

Habit of *Cyclea varians*

Genus:	
Albertisia	Liana; leaves penninerved; fruit with a style scar near the base.
Arcangelisia	Lianas with yellow wood; leaves penninerved; fruit with a lateral style scar.
Cissampelos	Scandent shrubs or lianas; leaves sometimes peltate, penninerved; inflorescence with large bracts; fruit with a style scar near the base.
Coscinium	Lianas or shrubs with yellow wood; leaves peltate, penninerved; fruit with a terminal style scar.
Cyclea	Woody climber; leaves often peltate, palmately nerved; fruit with a style scar near the base.
Fibraurea	Lianas with yellow wood and often white sap; leaves penninerved; fruit with a terminal style scar.
Haematocarpus	Lianas; leaves 3-veined from base; fruit with a style scar near the base.
Limacia	Woody climbers; leaves penninerved to 3-veined from base; fruit with a style scar near the base.
Pycnarrhena	Lianas or scandent shrubs; leaves usually penninerved; fruit with a style scar below the apex on the ventral side.
Stephania	Slender climbers; leaves peltate, palminerved; inflorescences umbelliform; fruit with a style scar near the base.
Tinomiscium	Woody climbers with white latex; leaves 3-veined from base, finely striated above; fruit with a terminal style scar.
Tinospora	Woody climbers with tuberculate stems; leaves penninerved; fruit with a terminal style scar.

Moraceae

Field characters:

Trees, shrubs or lianas, white sap usually present; leaves spirally arranged, sometimes alternate, simple; stipules present, often leaving a circular scar around stem; flowers unisexual, often grouped inside fleshy figs; ovary superior with 1–2 stigmas.

Description:

Habit trees, shrubs or lianas; often hemi-epiphytes, rarely with spines.

Sap usually present, white, sticky.

Stipules often large, forming a cap over the bud and leaving a circular scar after falling.

Leaves spirally arranged, sometimes alternate, rarely opposite; simple; margins entire to deeply lobed; venation pinnate, often conspicuous and/or 3-veined from the base.

Flowers unisexual, small; tepals 2–6 or absent; often enclosed inside a fleshy structure (figs); male flowers with 1–4 stamens, usually opposite tepals; female flowers with perianth often fused to ovary.

Ovary superior, stigmas (1 –)2.

Fruit (infructescence) drupe-like; sometimes dehiscent; often in complex, disc-like or globose structures (figs).

Confused with:

Urticaceae
Often herbs, sap absent; leaves glaucous below, cystoliths present; stigma 1.

Ulmaceae
Sap absent; stipules inconspicuous and soon falling off; flowers often bisexual and never condensed into heads.

Cannabaceae
Herbs, sap absent; leaves palmate; male flowers with 5 tepals and stamens; fruit a small nut or achene.

Magnoliaceae
Sap absent; flowers large, showy, solitary, bisexual.

Euphorbiaceae Usually sap absent; petiole swollen top and bottom; usually with 3 stigmas; fruit often dehiscing into 3 parts leaving a central columella.

Flacourtiaceae Usually sap absent; stipules inconspicuous; flowers often bisexual with many stamens, often with petals.

Artocarpus sp.

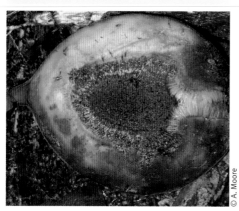

Cross-section of a fig

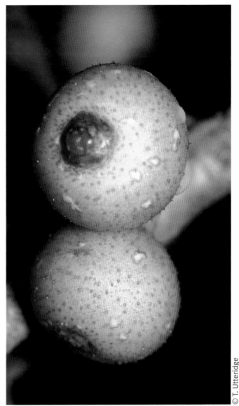

Prainea scandens

Small figs with sap

Major genera:

Antiaris	Trees; leaves hairy, margins entire; male flowers grouped into disc-like pedunculate inflorescences; female flowers solitary on short pedicels; fruit smooth.
Artocarpus	Trees; leaves margins entire to deeply lobed; stipule scars often circular; male and female inflorescences dense, globose stalked heads; 'fruit' globular, prickly, without basal bracts.
Ficus	Trees, shrubs or climbers; leaves margins entire to serrate, sometimes lobed; stipule scars circular; male and female flowers completely enclosed within a fig.
Parartocarpus	Trees; leaf margins entire; stipule scars never circular; 'fruit' a distinct involucre of 3–8 conspicuous basal bracts.
Prainea	Trees or climbers; leaves margins entire; scars not circular; mature drupes projecting out of the globular 'fruit' for most of their length.
Streblus	Trees or shrubs, mostly with spines; leaves margins entire to serrate; stipule scars not circular; male flowers grouped into many-flowered, long, narrow, 'catkin-like' inflorescences; female flowers solitary or in few-flowered racemes; fruit drupaceous with persistent tepals.

Myristicaceae

Field characters:

Trees with red sap; leaves alternate, simple, margins entire; stipules absent; fruit a leathery capsule with a large single suture, surface usually orange when mature, aril surrounding a single seed.

Description:

Habit trees, rarely shrub; twigs/branchlets often dark with a striate surface, cross-section often with rays in the wood.

Sap red sap from cut surfaces.

Stipules absent.

Leaves alternate (spirally or distichously arranged), simple, margins entire, often oblong or obovate, penninerved, glaucous below; hairs simple, stellate or T-shaped.

Flowers unisexual (plants dioecious or monoecious), very small, yellowish, white, pink or red; 3-merous, only 1 row of tepals; tepals inconspicuous, valvate, connate at base; filaments fused, often into a column.

Ovary superior.

Fruit fleshy to coriaceous, dehiscing into two parts along a single line of dehiscence; seed 1 (nutmeg) covered with red or orange aril (mace).

Confused with:

Annonaceae

Climbers and trees, sap absent; filaments free; fruit apocarpous.

Ebenaceae

Sap absent; flowers 4–5-merous, calyx accrescent; ovary 3- or more locular.

Lauraceae

Sap absent; anthers opening by valves; fruit indehiscent.

© RBG Kew

Inflorescence of *Knema* sp.

© RBG Kew

Flowers of *Knema* sp.

Horsfieldia sp.

Red sap

Ridges on the twigs

Fruit of *Horsfieldia* sp.

Genus:	
Endocomia	Leaves not glaucous below, brittle when dry; inflorescence non-woody, panicle-like, with basal bracts or their scars; flowers without a bracteole; aril entire or divided into segments to one-third of the way.
Gymnacranthera	Leaves glaucous below, not brittle when dry; inflorescence non-woody, panicle-like, with a basal bracts or their scars; flowers without a bracteole; aril divided into segments to near the base.
Horsfieldia	Leaves rarely glaucous below, brittle when dry; inflorescence non-woody, panicle-like, with basal bracts or their scars; flowers without a bracteole; aril entire or divided into segments to half way.
Knema	Leaves glaucous below, generally not brittle when dry; inflorescence woody, in fascicules, without basal bracts or their scars; flowers with a small bracteole; aril entire or shallowly lobed.
Myristica	Leaves often glaucous below, brittle when dry; inflorescence woody, panicle-like or in fascicules, sometimes with basal bracts or their scars; flowers with a bracteole; aril divided into segments to near the base.

Myrsinaceae

Field characters:

Leaves alternate or spirally arranged, simple, margins usually entire, penninerved; black dots or lines in the leaves, calyx, petals and on fruit; stipules absent; petals connate at the base, contorted; stamens opposite the petals; ovary 1-locular.

Description:

Habit trees, shrubs, lianas and herbs.

Sap rarely present.

Stipules absent.

Leaves spirally arranged or alternate, simple, margins usually entire, penninerved, often with red or black dots or lines.

Flowers usually 5-merous, usually with black dots or lines; calyx usually persistent; petals connate at base, contorted; stamens opposite the petals.

Ovary superior or semi-inferior, 1-locular.

Fruit a drupe, 1-locular, globose; calyx and style persistent.

Confused with:

Myristicaceae

Red sap present; leaves, calyx, petals and fruit without black dots; seed arillate.

Ochnaceae

Leaves, calyx, petals and fruit without black dots; stipules present.

Maesa macrocarpa

Fruit of *Ardisia* sp.

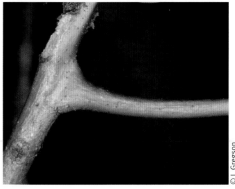

Branches swollen at base of *Ardisia* sp.

Flowers of *Ardisia* sp.

Habit of *Labisia* sp.

Genera:	
Aegiceras	Viviparous shrubs in mangroves; black dots on the leaves, calyx, petals and fruit; ovary superior; fruit woody, sickle-shaped.
Ardisia	Herbs, shrubs or small trees; branches swollen at the base, leaving large scars when shed; black dots on the leaves, calyx, petals and fruit; ovary superior.
Embelia	Climbers, sometimes with spines; black dots on the leaves, calyx, petals and fruit; ovary superior.
Labisia	Herbs with creeping stems; black dots on the leaves, calyx, petals and fruit; ovary superior.
Maesa	Small trees, shrubs and climbers; lines on the leaves, calyx, petals and fruit; ovary semi-inferior.
Rapanea	Small trees and shrubs; flowers unisexual, in clusters on short shoots; black dots in the leaves, calyx, petals and fruit; ovary superior.

Myrtaceae

Contributed by Eve Lucas

Field characters:

Trees and shrubs; leaves usually opposite, simple, margins entire, pellucid dots present (sometimes difficult to see); stipules absent; flowers with many stamens, petals free and imbricate; ovary (semi-)inferior.

Description:

Habit trees and shrubs; bark often flaky.

Sap absent.

Stipules absent.

Leaves opposite to spirally arranged, rarely in whorls, simple, margins entire, penninerved, often with a marginal vein; pellucid gland dots present (sometimes difficult to see), aromatic when crushed.

Flowers bisexual, petals free, mostly white, occasionally yellow or pinkish, imbricate, sometimes very small; stamens usually numerous.

Ovary inferior, rarely semi-inferior.

Fruit a berry or capsule.

Confused with:

Lecythidaceae
Leaves spirally arranged, without pellucid dots, margins often dentate.

Lythraceae
Leaves without pellucid dots; ovary superior.

Melastomataceae
Leaves 3- or 5-veined from the base, without pellucid dots; stamens few and with appendages.

Rubiaceae
Leaves without pellucid dots, intrapetiolar stipules present; corolla tubular.

Clusiaceae
Yellow or white exudates from cut tissue; leaves with black dots, not pellucid; ovary superior.

Bark of *Tristaniopsis* sp.

Inflorescences of *Melaleuca* sp.

Fruit of *Syzygium* sp.

© J. Gregson

Flower of *Syzygium* sp.

Major genera:

Decaspermum	Trees or shrubs; leaves opposite, sometimes with double intermarginal veins; inflorescence commonly a panicle; flowers silky pubescent; fruit fleshy.
Leptospermum	Shrubs; leaves alternate; fruit a round capsule splitting at top.
Melaleuca	Trees or shrubs; leaves spirally arranged, veins parallel to midvein; fruit a capsule.
Metrosideros	Trees; leaves opposite; fruit a round capsule splitting at the top.
Rhodamnia	Trees and shrubs; leaves opposite, 3-veined from base with whitish/iridescent hairs underneath; fruit fleshy.
Rhodomyrtus	Trees and shrubs, leaves opposite, 3-veined from base; fruit fleshy, sometimes moniliform.
Syzygium	Trees; leaves opposite, rarely spirally arranged, with a (double) intermarginal vein; inflorescence commonly a panicle or raceme; fruit fleshy.
Tristaniopsis	Trees with bark stripping off on long scrolls in some species; leaves alternate; flowers yellow; fruit a dry, round capsule splitting at top; seeds winged.
Xanthomyrtus	Mostly montane shrubs with small leaves and a dense habit; leaves opposite; flowers yellow, axillary, often in groups of 3; fruit fleshy.

Nepenthaceae

Field characters:

Herbs, often climbing; leaves spirally
arranged, simple, margins usually entire,
consisting of a blade, tendril, pitcher and lid.

Description:

Habit herb, often climbing, stems more-or-less woody.

Sap absent.

Stipules absent.

Leaves spirally arranged, simple, margins
usually entire, usually consisting of a blade,
tendril, pitcher and lid.

Flowers unisexual, regular, tepals 3–6.

Ovary superior.

Fruit a capsule, winged; seeds numerous, tiny.

Confused with:

Flagellaria
Leaves without pitchers.

Pitches of *Nepenthes* sp.

© S. Marsh

Flowers of *Nepenthes* sp.

© S. Marsh

Genus:

Nepenthes   As for the family.

Ochnaceae

Flowers of *Schuurmansia* sp.

Dried apocarpic fruit of *Brackenridgea palustris*

Dried leaf detail with close parallel veins of
Euthemis leucocarpa

Field characters:

Trees or shrubs; leaves spirally arranged, simple, margins entire to dentate; stipules present; flowers regular, sepals 5, petals 5–10; ovary superior, carpels usually free.

Description:

Habit trees to very small shrubs (almost herbs).

Sap absent.

Stipules present.

Leaves distichous or alternate, simple, margins entire to finely dentate; veins often fine and closely parallel.

Flowers bisexual, regular; sepals 5, free or connate at base, persistent; petals 5–10, free, soon falling off; stamens 5 or 10–many, anthers dehiscing lengthwise or shedding pollen through 1–2 apical pores.

Ovary superior, 2–10(–15) carpels, free or fused; style excentric.

Fruit a drupe, berry or capsule, sometimes with several on a torus; seeds small to large, sometimes winged.

Confused with:

Myrsinaceae
Black dots on the leaves, calyx, petals and fruit; stipules absent.

Genus:

Brackenridgea	Shrubs to trees; leaves distichous, margins entire; stipules free; flowers yellow or white; stamens >10, anthers with 2 longitudinal slits, fruit with 1–5 drupes per flower; 1 seed per drupe.
Euthemis	Shrubs; leaves alternate, margins toothed; flowers white or purple; stamens 5 with 1 apical pore per stamen; fruit a berry.
Ouratea (Gomphia)	Shrubs to trees; leaves distichous, margins toothed, with double intermarginal vein; stipules intrapetiolar fused; flowers yellow to white; stamens >10, anthers with 2 apical pores; fruit with 1–5 drupes per flower; 1 seed per drupe.
Sauvagesia (Neckia)	Small shrubs (almost herbs); leaves alternate, margins toothed; inflorescences axillary; flowers solitary, white; stamens 5 with longitudinal slits; fruit a capsule; seed not winged.
Schuurmansia	Shrubs with hollow branches; leaves alternate, margins entire to toothed; inflorescences terminal; flowers dark red; stamens 5 with longitudinal slits; fruit a capsule; seeds winged.

Oleaceae

Field characters:

Trees, shrubs and lianas, sap absent; leaves opposite, penninerved; stipules absent; flowers regular, corolla fused; stamens 2; ovary superior.

Description:

Habit trees, shrubs or lianas.

Sap absent.

Stipules absent.

Leaves (sub-)opposite, sometimes spirally arranged, simple or pinnate, penninerved, margins often entire, sometimes with dots; when dried, petioles often black and contrasting with the pale twig.

Flowers bisexual, regular, corolla usually 4-merous, fused; stamens 2.

Ovary superior, 2-celled, 1 or 2 ovules per cell.

Fruit a drupe, berry or capsule.

Confused with:

Apocynaceae
White sap present; stamens 5; fruit in pairs.

Labiatae
Stamens (2–)4 or 5, flowers usually zygomorphic.

Loganiaceae
Stipules or an interpetiolar ridge present; stamens 4 or 5.

Rubiaceae
Interpetiolar stipules present; stamens more then 2; ovary usually inferior.

Habit and fruit of *Jasminum* sp.

© R. de Kok

White twig and black petioles of *Ligustrum glomeratum*

© R. de Kok

Flower of *Chionanthus* sp.

© T. Utteridge

Genera:

Chionanthus	Trees or shrubs; leaves simple, sometimes with dots; inflorescences usually axillary; fruit drupaceous, often black and glaucous.
Jasminum	Climbers or shrubs; leaves sometimes spirally arranged, simple or pinnate; fruit a deeply 2-lobed berry with persistent calyx.
Myxopyrum	Climbers; leaves simple, thick coriaceous, with 3-longitudinal veins, glandular below; fruit drupaceous, globose.

Orchidaceae

Field characters:

Herbs, sometime climbing, not aromatic; leaves spirally arranged, simple; flowers zygomorphic, 3-merous; stamens adnate to the style forming a column; ovary inferior; seeds many, tiny.

Description:

Habit herbs, often epiphytic, sometimes climbing, sometimes saprophytic; stems sometimes swollen into pseudo-bulbs, roots with a velvet sheath.

Sap absent.

Stipules absent.

Leaves spirally arranged, rarely opposite or whorled, simple, sometimes reduced to scales, margin entire, parallel-veined; sheathing at base, sheath usually closed.

Flowers bisexual, rarely unisexual, usually zygomorphic, rarely actinomorphic, usually resupinate; median petal (lip) very different in shape, size and coloration from the lateral petals; stamens adnate to the style forming a column; rarely stamens free, 1(– 3); pollen coherent into a small waxy body.

Ovary inferior.

Fruit a dry capsule, splitting lengthways; seeds numerous, tiny.

Confused with:

Amaryllidaceae
Flowers actinomorphic, with >3 stamens.

Burmanniaceae
Flowers with a fused corolla.

Corsiaceae
Flowers with >3 stamens.

Hypoxidaceae
Flowers actinomorphic, with >3 stamens.

Costaceae
Ligule a forming a ring above the pseudo-petiole; stamens 1, free.

Zingiberaceae
Roots aromatic; ligule a small leaf-like structure at base of leaf-blade; stamens 1, free.

© S. Marsh

Dendrobium leonis

© J. Gregson

Coelogyne (section *Rigidiformes*) aff. *rigidiformis*

Genera:

Many genera of Orchidaceae, too numerous to list in this non-specialist field guide, are found in East Sabah.

Palmae

Contributed by Bill Baker

Field characters:

Trees, shrubs or climbers; leaves compound (pinnate, palmate or bipinnate), rarely entire, often armed, with sheathing base; inflorescences spicate or paniculate, with at least 1 conspicuous bract; bract(s) persistent or caducous; flowers small, regular, 3-merous; ovary superior.

Description:

Habit trees, shrubs and climbers; stems cylindrical and marked with conspicuous ring-like leaf scars, usually not branched.

Sap absent.

Stipules absent.

Leaves spirally arranged, compound (pinnate, palmate or bipinnate), more rarely entire, often armed with spines, with sheathing base (sometimes forming crownshaft); leaflets/segments/lamina folded (plicate), margins not entire.

Inflorescences a spike or panicle, with 1–many conspicuous bracts; bracts sometimes falling prior to anthesis, presented above, among or below the leaves.

Flower often unisexual, small, 3-merous, often arranged in clusters; stamens usually 6, but sometimes many more.

Ovary superior, 1–3-locular, 1 ovule per locule.

Fruit typically a 1-seeded berry or drupe, sometimes with up to 3 seeds, sometimes covered with scales resembling snake skin.

Confused with:

Pandanaceae
Leaves linear, never compound, usually 3-ranked, margins spiny; flowers without perianth.

Cycadaceae
Leaves with circinate venation, not plicate; cone-bearing.

© R. de Kok

Fruit and flowers of *Iguanura* sp.

Major genera:	
Areca	Small trees or shrubs, unarmed; leaves pinnate or entire; female inflorescences below the leaves; flowers occurring only at the base of inflorescence branches.
Arenga	Moderate to robust shrubs and trees, unarmed; leaves pinnate, sheaths fibrous, leaflet base asymmetrical and v-shaped in section.
Borassodendron	Robust tree; leaves palmate with wedge-shaped segments; petiole not armed with spines but sharp.
Caryota	Trees and shrubs, unarmed, leaves bipinnate.
Calamus	Climber (rattan), rarely shrub-like, spiny; leaves in climbers with a knee-like bulge or wrinkle on sheath below petiole base, leaf tip a climbing whip (cirrus) or leaf sheath a climbing whip (flagellum); inflorescences with major bracts tubular and persistent.
Ceratolobus	Climber (rattan), spiny; leaves with a knee-like bulge or wrinkle on sheath below petiole base, leaf tip a climbing whip (cirrus), sometimes diamond-shaped leaf; inflorescence at anthesis completely enclosed within a pod-like boat-shaped bract.
Daemonorops	Climber (rattan), rarely shrub-like, spiny; leaves in climbers with a knee-like bulge or wrinkle on sheath below petiole base, leaf tip a climbing whip (cirrus) or leaf sheath a climbing whip (flagellum); inflorescences with major bracts tubular but splitting longitudinally and sometimes falling.
Iguanura	Small trees or shrubs, unarmed; leaves pinnate or entire, leaflets with jagged distal margin; inflorescences among or below the leaves.
Korthalsia	Climber (rattan), spiny; leaves lacking a knee-like bulge or wrinkle on sheath below petiole base, leaf tip a climbing whip (cirrus), leaflets diamond-shaped with jagged distal margins; flowering above the leaves; stems dying after flowering.
Licuala	Shrub to small tree; leaves palmate with wedge-shaped segments, sometimes entire, petiole armed with spines.
Oncosperma	Large tree; leaves pinnate, spines abundant on trunk, leaf and inflorescence.
Pholidocarpus	Robust tree; leaves palmate with wedge-shaped segments, petiole with conspicuous spines.
Pinanga	Small trees or shrubs, unarmed; leaves pinnate or entire; inflorescences below the leaves, female flowers present along whole length of inflorescence branches.
Plectocomia	Climber (rattan), spiny; leaves lacking knee-like bulge or wrinkle on sheath below petiole base, leaf tip a climbing whip (cirrus); first-order branches of inflorescences pendulous and with conspicuous bracts; stems dying after flowering.
Plectocomiopsis	Climber (rattan), spiny, lacking knee-like bulge or wrinkle on sheath below petiole base, leaf tip climbing whip (cirrus), conspicuous ocrea present, stems dying after flowering.
Retispatha	Climber (rattan) or short-stemmed palm, spiny, lacking any climbing organs; inflorescences with net-like bracts covering all branches.
Salacca	Stemless shrub; leaves spiny pinnate; inflorescences emerging through the back of the leaf sheath.

Pandanaceae

Contributed by Bill Baker

Field characters:

Trees, shrubs or climbers, dioecious; leaves spirally arranged, simple, linear, spiny; stipules absent; inflorescences comprising 1–many compact heads, usually covered by bracts; flowers small, perianth absent, stamens numerous; fruiting heads consisting of many smaller fruitlets.

Description:

Habit trees, shrubs and root climbers, dioecious; stems branching; sometimes with stilt roots.

Sap absent.

Stipules absent, but climbing species with membranous ligules flanking leaf base.

Leaves spirally arranged, usually 3-ranked, simple, linear, margins usually with spines.

Inflorescences comprising 1–many compact heads; bracts conspicuous, fleshy and often colourful and fragrant.

Flower unisexual, small, perianth absent; stamens numerous.

Ovary superior.

Fruit fibrous to fleshy drupes or berries, consisting of 1–several carpels, 1–many seeded, aggregated in compact fruiting heads.

Confused with:

Palmae
Leaves usually compound, never linear; flowers with perianth.

Ruscaceae (Dracaena)
Leaves not spiny; flowers large with distinct perianth and not presented in compact heads.

© T. Utteridge

Freycinetia sp.

© J. Dransfield

Pandanus calcicolus

Genera:

Freycinetia	Climber; membranous ligules flanking leaf base.
Pandanus	Shrubs and trees; stilt roots sometimes present.

Piperaceae

Field characters:

Herbs to shrubs, stems swollen at the nodes; leaves simple, margins entire; inflorescences spicate, opposite the leaves; flowers small, without perianth; ovary superior.

Description:

Habit herbs to shrubs or small trees, rarely climbing; stems swollen at the nodes.

Sap absent.

Stipules usually present, sometimes adnate to petiole or modified to auricles.

Leaves usually alternate; simple, margins entire, 3-veined from base, often dotted with glands.

Inflorescences spicate or racemes, usually opposite the leaves, axillary or extra-axillary.

Flowers small; unisexual or bisexual; perianth absent; stamens 1–10.

Ovary superior, 1-celled, 1 ovule.

Fruit a berry.

Confused with:

Chloranthaceae

Leaves opposite; margins toothed; inflorescences a compound spike; ovary inferior.

Urticaceae

Stems not swollen at the nodes; leaves with cystoliths; perianth present; stipules present.

Fruit of *Piper* sp.

Habit of *Piper* sp.

Genera:

Peperomia	Herbs; stipules absent.
Piper	Shrubs or trees, rarely climbing; stipules present.

Poaceae

Field characters:

Herbs, sometimes woody; stems hollow, apart from the solid joints; leaves 2-ranked, ligules present; flowers small.

Description:

Habit annual or perennial herbs, sometimes woody, sometimes climbing, sometimes with rhizomes or vegetative parts above or below ground shoots; stems rounded, rarely flattened in cross-section, usually hollow, solid nodes present.

Sap absent.

Stipules absent.

Leaves 2-ranked, usually linear, without a petiole, ligules present.

Flowers small, surrounded by bracts (lemma and palea), organised in small inflorescences (spikelets); without a perianth, stamens 3, rarely 6; utricle (perygynium) absent.

Ovary superior.

Fruit caryopsis (a dry fruit like a rice grain), rarely berry-like.

Confused with:

Cyperaceae

Stems solid, without joints, leaves often 3-ranked; flowers subtended by bracts; arranged in spikelets.

Juncaceae

Stems solid to hollow, without joints; leaves spirally arranged, without a ligule; flowers not subtended by bracts.

© R. de Kok

Swollen node of *Poaceae* stem

Stems hollow, round in cross-section, with leaves in 2-ranks

© R. de Kok

Poaceae habit

Major genera:

Many genera of Poaceae, too numerous to list in this non-specialist field guide, are found in East Sabah.

Podocarpaceae

Field characters:

Trees; leaves simple, entire; flowers in an unisexual cone (a condensed sort of inflorescence).

Description:

Habit trees.

Sap absent.

Stipules absent.

Leaves opposite to spirally arranged, simple, margins entire, needle-like to broad-leaved, or scale-like, venation appearing parallel.

Flowers in unisexual cones, female flowers simple.

Ovary superior.

Fruit a berry on a swollen receptacle.

Confused with:

Casuarinaceae

Twigs jointed, flowers with perianth.

Podocarpus motleyi

© J. Dransfield

Dried cone of *Nageia wallichima*

© R. de Kok

Fruit of *Podocarpus neriifolius* with swollen receptable

© T. Utteridge

Genera:	
Dacrydium	Leaves spirally arranged, needle-like, a single leaf type present on twigs.
Dacrycarpus	Leaves spirally arranged, scales to needle-like, two different leaf types present on twigs.
Falcatifolium	Leaves spirally arranged, scales-like or broad (>2 mm wide) and bilaterally flattened.
Nageia	Leaves opposite, broad (>2 mm wide) and bifacially flattened.
Phyllocladus	Twigs flat and resembling leaves.
Podocarpus	Leaves spirally arranged, broad (>2 mm wide) and bifacially flattened, sometimes with dots.

Polygalaceae

Field characters:

Herbs to trees, sometimes climbers; leaves simple, margins entire; stipules absent; flowers 5-merous, bisexual; ovary superior.

Description:

Habit herbs to trees, rarely climbers, sometimes saprophytic herbs.

Sap absent.

Stipules absent.

Leaves spirally arranged, rarely sub-opposite, simple, margins entire, sometimes with scattered glands on lower surface; rarely saprophytic herbs with scale-like leaves without any chlorophyll.

Flowers bisexual, 5-merous, usually zygomorphic, sepals free; petals 3 or 5, free or connate; stamens 2–10, connate.

Ovary superior, 1–8-locular; style not always persistent, usually articulate with the fruit.

Fruit a berry, capsule, samara or drupe.

Confused with:

Leguminosae (Caesalpinioideae, Mimosoideae and Papilionoideae)
Leaves often compound, leaves or leaflets with pulvini; stipules usually present; ovary 1-locular.

Trigoniaceae
Leaves simple, margins entire, lower surface covered with matted white simple hairs; stamens 6, forming a tube; fruit 3-winged.

Flowers of *Polygala* sp.

Fruit of *Xanthophyllum flavescens*

Epirixanthes sp.

© J. Gregson

Habit of *Xanthophyllum flavescens*

Genera:	
Epirixanthes	Saprophytic herbs; leaves scale-like, not green; stamens 2–5; fruit a indehiscent drupe or berry, not winged.
Polygala	Herbs or shrubs, rarely climbers; leaves green; inflorescence sometimes supra-axillary; stamens (6–)8; fruit a winged capsule.
Salomonia	Herbs; leaves green, 3-veined from base; stamens 4–6; fruit a dehiscent capsule, not winged.
Securidaca	Climbers, glands at nodes; leaves green; stamens 8; fruit a samara.
Xanthophyllum	Trees or shrubs, sometimes with spines; leaves green, usually drying yellow, sometimes with glands; fruit indehiscent, drupe or berry-like, often with a hairy band below the calyx.

Proteaceae

Field characters:

Small trees and shrubs; leaves alternate, rarely opposite; stipules absent; inflorescences usually complex; flowers 4-merous, 1 whorl of tepals; stamens 4.

Description:

Habit trees or shrubs.

Sap absent.

Stipules absent.

Leaves alternate or rarely sub-opposite, simple, sometimes lobed, margins entire or toothed.

Flowers usually bisexual; 1 whorl of tepals, tubular when young, splitting and reflexing at maturity; tepals often brightly coloured, 4-merous; stamens 4, opposite the tepals lobes.

Ovary superior, 1-locular, often with a persistent 'style' (pollen presenter).

Fruit a nut or drupe-like; seeds 1(–2).

Confused with:

Loranthaceae

Parasitic epiphytic shrubs; leaves usually opposite; ovary inferior.

Aralidiaceae

Trees, shrubs or climbers; petioles clasping; inflorescence umbelliform; ovary inferior; fruit a fleshy berry.

Flowers of *Helicia* sp.

© T. Utteridge

Fruit with pollen presenter of *Helicia* sp.

© R. de Kok

Fruit of *Heliciopsis artocarpoides*

© J. Gregson

Genera:

Helicia	Leaves not lobed; flowers bisexual.
Heliciopsis	Juvenile leaves large, deeply pinnatifid; flowers unisexual.

Rhamnaceae

Field characters:

Trees, shrubs or lianas; leaves alternate, simple, venation sclariform; stipules present, but usually small; flowers regular, stamens opposite the free petals, intra-staminal disk present.

Description:

Habit trees, shrubs or lianas; lianas often with tendrils, often with straight or curved spines.

Sap absent.

Stipules present, but usually small; sometimes formed into spines.

Leaves alternate to rarely opposite, simple, margins entire to serrate; venation pinnate, 3-nerved or palminerved, with tertiary scalariform venation.

Flowers bisexual or unisexual, regular, usually 4–5-merous; hypanthium resembling a calyx-tube; petals small, free, usually enclosing the stamens; stamens opposite the petals; intra-staminal disk present.

Ovary inferior to superior.

Fruit usually a drupe, sometimes a capsule; often winged.

Confused with:

Celastraceae
Leaves usually (sub-)opposite; stamens opposite the sepals.

Euphorbiaceae
Flowers unisexual; stamens not opposite the petals.

© J. Dransfield

Flower of *Ziziphus horsfieldii*

Genera:

Colubrina	Trees or shrubs; leaves pinnately veined; ovary superior; fruit a 3-locular capsule.
Gouania	Climber, often with tendrils; leaves 3–5-nerved; ovary inferior; fruit a 3-winged capsule.
Rhamnus	Trees or shrubs, rarely climbing, often with spines; leaves pinnately veined; ovary superior; fruit a drupe.
Smythea	Climber with hooks; leaves penninerved, transversely parallel-veined; ovary superior; fruit a dehiscing samara, with a small apical wing.
Ventilago	Climber with hooks; leaves penninerved, transversely parallel-veined; ovary semi-inferior; fruit an indehiscent samara, with a long wing.
Ziziphus	Trees or scramblers with prickles; leaves 3-nerved; ovary semi-inferior; fruit a 1-seeded drupe.

Rosaceae

Field characters:

Usually trees; leaves alternate, rarely opposite or whorled, usually simple, margins entire, toothed or pinnately lobed; stipules present, soon falling; flowers regular, hypanthium present, tubular or cupular or enclosing the fruit; style terminal.

Description:

Habit trees, shrubs or herbs, sometimes with thorns.

Sap absent.

Stipules usually present, falling off early.

Leaves alternate, simple to compound, margins entire or toothed; often with two distinct glands on the petiole or on the lower leaf blade.

Flowers bisexual, mostly regular, 5-merous with a distinct calyx and corolla; hypanthium present; stamens usually numerous.

Ovary inferior to superior, style terminal.

Fruit a drupe or a head of drupelets.

Confused with:

Chrysobalanaceae
Always trees; flowers more-or-less zygomorphic, style more-or-less excentric.

Symplocaceae
Trees or shrubs; leaves drying bright yellow green; stipules absent; ovary inferior.

Saxifragaceae
Trees or shrubs, rarely climbers or herbs; usually without stipules.

© M. Coode

Prunus sp.

© T. Utteridge

Rubus sp.

Major genera::

Prunus	Trees; leaves simple, thorns absent, often with 2 distinct glands on the petiole or on the lower lamina; ovary superior; fruit a drupe.
Rubus	Herbs to shrubs, sometimes climbing; leaves simple (can be lobed), usually with thorns; ovary superior; fruit consist of many small drupes.

Rubiaceae

Contributed by Aaron P. Davis

Field characters:

Leaves simple, opposite, margins entire; interpetiolar stipules present; flowers actinomorphic, corolla tubular; stamens the same in number as the corolla lobes; ovary inferior, rarely superior.

Description:

Habit trees, shrubs, climbers, herbs, and epiphytes.

Sap absent.

Stipules interpetiolar stipules present and usually conspicuous.

Leaves opposite or verticillate, sometimes 1 of the pair reduced; simple, margin entire.

Flowers bisexual or unisexual; corolla actinomorphic, tubular; stamens the same in number as the corolla lobes; anthers fixed to the corolla tube.

Ovary inferior, rarely superior.

Fruit indehiscent (berries or drupes) or dehiscent (capsules or mericarps), sometimes united and fused into a syncarp, rarely a multi-seeded stone; endosperm present and usually conspicuous.

Confused with:

Apocynaceae
White exudate present, stipules absent; ovary superior; fruit paired.

Caprifoliaceae
Leaf margins usually serrate; stipules absent.

Loganiaceae
Stipules absent; ovary superior.

Rhizophoraceae
Leaf margins usually serrate; petals free; stamens twice as many as corolla lobes.

© RBG Kew

Argostemma sp.

© J. Gregson

Lasianthus sp.

Rubiaceae cont.

Major genera:	
Herbs	
Acranthera	Corolla large; fruit fleshy, indehiscent, narrowly cylindrical; seeds numerous, small, surface pitted.
Argostemma	Leaf pairs usually anisophyllous; leaves in pairs of the same size or one sometimes absent; fruit capsular, opening by an apical operculum, crowned by a persistent calyx; seeds numerous, small.
Geophila	Creeping and rooting at nodes; leaves orbicular to reniform; fruit a red fleshy indehiscent drupe; 2 single-seeded pyrenes.
Hedyotis	Stipules fimbriate; fruit usually septicidally dehiscent; seeds numerous, small.
Ophiorrhiza	Inflorescence a helicoid or scorpoid cyme; fruit loculicidally dehiscent, laterally compressed, obtriangular to heart-shaped in outline; seeds numerous, small.
Spermacoce	Stipules entire or fimbriate; fruit a 2-valved capsule, dehiscing from apex downwards; seeds 2 (1 in each half of the fruit), rather large.
Climbers	
Paired, curved thorns (hooks) present	
Oxyceros	Fruit berry-like, fleshy-leathery, indehiscent; seeds numerous, lens-shaped and embedded in a pulpy matrix.
Uncaria	Fruit condensed into a spherical head, but not fused, loculicidally dehiscent; seeds numerous, compressed and winged.
Thorns absent	
Morinda	Trees or shrubs or climbers; inflorescence usually terminal, flowers fused into a head; fruit fused forming a rather fleshy syncarp, each fruit unit containing a single-seeded pyrene; numerous seeds in each syncarp, but only 1 in each fruit unit.
Mussaenda	Trees, scrambling or loosely twining; inflorescence terminal with 1 or more calyx lobes of each inflorescence enlarged into a bright (usually yellow, orange or white) show bract (semaphyll); corolla lobes valvate-reduplicate in bud (corolla lobes folded or ridged); fruit free, berry-like, indehiscent, fruit wall rather tough and leathery; seeds numerous, minute, brown to black.
Paederia	Plants climbing or scrambling by means of winding stems; foetid smelling, especially when bruised; fruit free, dehiscent, containing 2 single-seeded, winged pyrenes.
Schradera	Climbing or scrambling by means of internodal adhesive adventitious roots; flowers condensed into a flowering head; fruit tightly packed but separate, berry-like (indehiscent) and rather fleshy; seeds numerous, small.
Trees, treelets and shrubs	
Flowers and fruit in tight spherical heads	
Ludekia borneensis	Trees; stipules fused (later free) into a cone; flowering heads small and usually >5; flowers tightly packed but not fused; stigma obovoid with 7–9-longitudinal ridges; fruiting heads with free fruitlets.
Morinda	See 'Climbers'.
Myrmeconauclea	Shrubs, usually close to rivers and streams, ant chambers usually present; stipules flattened, elliptic to ovate-oblong; young flowering

Rubiaceae cont.

Major genera cont.

	heads not enclosed by bracts; flowers tightly packed but not fused; stigma globose; fruiting heads with fruitlets partially fused (upper part of the hypanthium only) into a fleshy pseudosyncarp, which falls apart with age.
Nauclea	Trees; stipules flattened, ovate, elliptic or obovate; young flowering heads not enclosed by bracts; flowers tightly packed and (hypanthia) fused together; stigma spindle-shaped; fruiting heads with fruitlets fused into a fleshy syncarp.
Neonauclea	Trees; stipules flattened, ovate, elliptic or obovate; young flowering heads enclosed by a pair of large, deciduous, opposite bracts; flowers tightly packed but not fused; stigma globose to ovoid; fruiting heads with free (rather stiff and loose) fruitlets.
Neolamarckia cadamba	Trees to 40 m; stipules narrowly triangular; young flowering heads not enclosed by bracts; flowers tightly packed but not fused; stigma spindle-shaped; fruiting heads with free (rather fleshy) fruitlets.

Inflorescences distinctly paired-axillary, in several nodes per inflorescence and not in heads; fruit fleshy and indehiscent

Corolla lobes overlapping to the left when in bud (corolla contorted)

Aidia	Plants often drying blackish; inflorescences appearing axillary but actually terminal and opposite a single leaf; each fertile node separated by a node with a normal leaf pair; stipules triangular; fruit indehiscent, globose to ellipsoid, rather small (<1 cm in diam.), fleshy to fleshy-leathery, 2-locular, each locule containing several small angular seeds in a pulpy matrix.

Mussaenda sp.

Rubiaceae cont.

Major genera cont.	
Cowiea borneensis	Inflorescence spicate (unbranched), very long and slender and pendulous; flowers 5-merous; fruit black; seeds with a fingerprint-like pattern.
Hypobathrum	Inflorescence raceme-like (shortly branched) slender or condensed (numerous cymes); flowers usually 4-merous; fruit usually pink; seeds with a fingerprint-like pattern.
Diplospora	Stipules distinctly long-awned; inflorescence a condensed, more-or-less sessile, few to many-flowered cyme or fascicle, subtended by 4-lobed calyx-like bracts; flowers 4-merous; fruit usually red or orange; seeds smooth or with a shallow groove.

Corolla lobes not overlapping when in bud (corolla valvate)

Timonius	Flowers unisexual; terminal portions of shoots and fertile parts often densely covered with soft, fine, brown hairs; inflorescence restricted to the terminal portions of the shoot, few to many flowered pedunculate cymes; fruit with several to many locules, with 1 pyrene in each locule.
Jackiopsis ornata	Tree 10–12 m (or taller); flowers hermaphroditic; inflorescence restricted to the terminal portions of the shoot; stipules connate into a tubular sheath, 8–15 awned; calyx lobes 5, 3 enlarged and obvious and 2 reduced; fruit surmounted by 3 persistent, enlarged calyx lobes; seeds 1 (by abortion).
Lasianthus	Flowers bisexual; inflorescence a sessile, rarely pedunculate cluster of few to several flowers, sometimes with large bracts; hairs in corolla throat rather soft, white or whitish; stigma 2-lobed; fruit 3–9-locular, usually ripening blue, with 1 pyrene in each locule.
Maschalocorymbus	Flowers unisexual or functionally so; inflorescence trichotomously corymbose, many flowered; hairs in corolla throat stiff; stigma usually >3-lobed; fruit 5-locular; seeds numerous, very small.
Pleiocarpidia	Flowers unisexual or functionally so; inflorescences paniculate or trichotomously corymbose, many flowered; hairs in corolla throat soft; stigma usually >3-lobed; fruit 5–8-locular; seeds numerous, very small.
Praravinia	Flowers unisexual or functionally so; inflorescence with 2 involucels, with up to 13-fascicled flowers; calyx lobes always fewer in number than the corolla lobes; hairs in corolla throat stiff; stigma usually >3-lobed; fruit 5–16-locular; seeds numerous, very small.
Urophyllum	Flowers unisexual or functionally so; inflorescence umbellate or 1-flowered; hairs in corolla throat soft; stigma usually >3-lobed; stigmata cohering into a flat disk; fruit 5–16-locular; seeds, numerous, very small.
Psydrax	Often substantial trees; flowers bisexual; inflorescence usually a many-flowered umbel-like cyme; hairs in corolla throat soft; stigma (pollen presenter) distinctly exserted and knob-like; fruit 2-locular, with 1 pyrene in each locule.
Streblosa	Inflorescence terminal but becoming 'axillary' and sometimes paired, pedunculate; flowers bisexual; flowers often condensed; flowers hermaphroditic; hairs in corolla throat soft; stigma 2-lobed; fruit 2-locular, splitting in 2, with 1 pyrene in each locule.

Rubiaceae cont.

Uncaria sp.

© J. Gregson

Rubiaceae cont.

Major genera cont.

Inflorescence terminal or more-or-less so, rarely axillary; fruit fleshy and indehiscent

Corolla lobes overlapping when in bud (corolla contorted)

Flowers and fruit generally small (<1.5 cm wide), each seed with a distinct hole (hilar cavity) in its dorsal surface

Pavetta	Leaves with bacteria nodules; flowers 4-merous; style club-shaped; fruit shiny and minutely wrinkled when dry; seeds 1 or 2.
Tarenna	Stipules often drying blackish; flowers 5-merous, rarely 4-merous; style club-shaped; fruit shiny and minutely wrinkled when dry; seeds 3–several.
Ixora	Stipules distinctly long-awned, drying the same colour as the stems; petioles usually with a distinct abscission between petiole and stem; flowers 4-merous; style distinctly 2-lobed; fruit dull and smooth when dry; seeds 2, rarely 1.

Flowers and/or fruit generally large (>2 cm and mostly >4 cm); seeds embedded in a placental pulp

Gardenia	Parts coated in a clear to translucent, sticky exudate (appearing shiny when dried); stipules forming a short tubular sheath around the stem, apex becoming truncate; corolla usually >4 cm long; fruit 1-locular; placentas attached to the edge of the inner fruit wall (parietal).
Porterandia	Leaf pairs anisophyllous; corolla usually <3 cm long, plain; fruit 2-locular; placentas attached to the middle of the fruit (axile).
Rothmannia	Flowers and fruit subtended by 3 leaves; corolla usually >4 cm long, often with darker spots or markings in the throat; fruit 1-locular; placentas attached to the edge of the inner fruit wall (parietal).

Corolla lobes not overlapping (valvate or valvate-reduplicate) when in bud

Seeds (pyrenes) 2, or rarely 1

Chassalia	Inflorescence branches usually highly coloured (white, red or purple), and often rather succulent; stipules white-papery on dried specimens; corolla rather long and curved; fruit fleshy-succulent, containing 2 single-seeded pyrenes; pyrenes smooth or few ridged on dorsal surface, with a distinct internal cavity.
Psychotria	Inflorescence branches usually green or brown, more-or-less woody; corolla short and straight; fruit containing 2 single-seeded pyrenes; pyrenes distinctly ridged or rarely smooth on dorsal surface, lacking a distinct internal cavity.
Streblosa	See 'Inflorescences distinctly paired-axillary'.
Gaertnera	Stipules forming a more-or-less tubular sheath around the stem; corolla short and straight; fruit superior, containing 2 single-seeded pyrenes; pyrenes smooth on dorsal surface, more-or-less flat on ventral surface.
Saprosma	Plants foetid when bruised; stipules rather woody, 2–5-awned at the apex; corolla short and straight; fruit fleshy-succulent, containing 2 single-seeded pyrenes; pyrenes smooth on dorsal surface.
Prismatomeris	Terminal portions of shoot with a rather distinct median longitudinal ridge; shoots appearing shiny and 'varnished' when dry; corolla long, straight or sometimes very slightly curved; fruit containing 2 (hemispherical) or 1 (almost spherical) seed(s); seeds with a large and obvious cavity hole on the dorsal face.

Rubiaceae cont.

Major genera cont.

Fruit indehiscent; seeds numerous, brown or black

Mycetia	Terminal portions of shoot smooth and shiny, becoming distinctly corky (and easily peeling) when dried; corolla lobes valvate in bud (corolla lobes flat when open); fruit wall soft and watery-succulent.
Mussaenda	See 'Climbers'.

© J. Gregson

Urophyllum sp.

Rutaceae

Field characters:

Trees, shrubs and lianas; leaves often compound, with pellucid gland dots, bruised leaves with an orange/citrus smell; stipules lacking; flowers with a disk.

Description:

Habit trees, shrubs and sometimes lianas, rarely herbs; often with thorns or spines.

Sap absent.

Stipules absent.

Leaves opposite or spirally arranged, with pellucid gland dots, bruised leaves with an orange/citrus smell; usually pinnately or palmately compound, or if unifoliolate with the petiole jointed; margins entire to serrate; petioles sometimes winged, sometimes with a swollen apex.

Flowers usually bisexual, 5-merous, regular, sweet smelling, often white, pale yellow or pale pink; petals usually free, with pellucid gland dots; stamens 2–many, arranged in 2 whorls, filaments thick, sometimes flattened, sometimes connate at the base.

Ovary superior, with (1–)4–5(–many) locules; with a conspicuous disc at the base of the ovary.

Fruit a dehiscent capsule, follicle, berry or drupe, often strongly lobed, with pellucid gland dots; seed often black and shiny.

Confused with:

Burseraceae
Plants resinous; leaves without pellucid dots.

Meliaceae
Leaves rarely with pellucid dots; stamens united to form a tube.

Simaroubaceae
Leaves without pellucid dots; carpels usually free.

Capsules with shiny seeds of *Euodia latifolia*

© R. de Kok

Habit of *Zanthoxylum integrifoliolum*

© D. Hicks

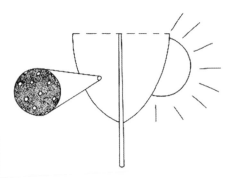

Leaves with pellucid gland dots

© RBG Kew

© T. Utteridge

Habit of *Melicope* sp.

Major genera:	
Acronychia	Trees and shrubs; leaves opposite, unifoliolate; fruit a drupe.
Citrus	Small trees or shrubs; leaves unifoliolate, spiny, petiole often winged; stamens fused; fruit fleshy, >4 cm diameter.
Clausena	Trees; leaves alternate, imparipinnate, leaflets alternate; inflorescence a panicle; fruit a berry.
Glycosmis	Shrubs; leaves alternate, unifoliolate or usually imparipinnate, rarely 3-foliolate, leaves often dry yellow; leaflets alternate; fruit a dry or juicy berry.
Luvunga	Climbers, stems sometimes spiny; leaves alternate, 3-foliolate; fruit a glandular berry.
Maclurodendron	Trees or shrubs; leaves opposite, unifoliolate; fruit a drupe.
Melicope	Small trees; leaves opposite, usually palmately compound; fruit a capsule.
Micromelum	Shrubs; leaves alternate, imparipinnate; fruit a berry.
Murraya	Trees or shrubs; leaves alternate, imparipinnate, leaflets alternate; fruit a small berry with mucilaginous pulp.
Pleiospermium	Trees to shrubs usually spiny; leaves alternate, unifoliolate, petiole winged; fruit a berry with a rough-glandular pericarp.
Tetractomia	Trees or shrubs; leaves opposite, unifoliolate, margins entire, leathery; fruit capsule with winged seeds.
Zanthoxylum	Trees, shrubs and climbers, stem spiny; leaves alternate, pinnate or 3-foliolate, rarely unifoliolate, sometimes imparipinnate; leaflets opposite; stamens with appendages; fruit 1–5 free or fused at base follicles.

Santalaceae

Field characters:

Trees, shrubs or climbers; leaves simple, margins entire; stipules absent; flowers regular; stamens the same in number as, and opposite, the perianth segments; ovary usually inferior.

Description:

Habit small to medium-sized trees, shrubs or climbers; sometimes with spines; parasitic on the roots of other woody plants.

Sap absent.

Stipules absent.

Leaves alternate, rarely opposite, simple, margins entire, often reduced to a scale-like structure; venation usually sub-palmate; often turning black when drying.

Flowers bisexual, small, regular, 3–5-merous, with only 1 perianth whorl, stamens the same in number as, and opposite, the perianth segments.

Ovary inferior, rarely superior.

Fruit drupaceous.

Confused with:

Celastraceae
Leaves usually opposite; stipules present; flowers with 2 perianth whorls (calyx and corolla); ovary superior.

Loranthaceae
Parasitic on stems and branches; corolla well developed; ovary inferior.

Olacaceae
Two perianth whorls present (calyx and corolla); ovary usually superior.

Opiliaceae
Ovary superior.

Viscaceae
Parasitic on stems and branches; tepals tiny; ovary inferior.

Fruit of *Scleropyrum* sp.

© J. Dransfield

Flowers of *Scleropyrum pentandrum*

© T. Utteridge

Genera:

Dendromyza	Climber with twining stems; leaves sometimes scale-like; ovary inferior; fruit with 1 seed.
Dendrotrophe	Climber without twining stems; ovary inferior; fruit with many seeds.
Scleropyrum	Trees with usually branched spines on stem; leaves alternate; ovary inferior; fruit without a swollen receptacle.

Sapindaceae

Field characters:

Trees, shrubs or rarely herbaceous climbers, sap absent; leaves spirally arranged, usually compound, usually paripinnate, often ending in a free rachis tip; stipules absent; stamens usually 8, free; ovary superior.

Description:

Habit trees and shrubs or rarely herbaceous climbers.

Sap not present or very little.

Stipules absent (sometimes pseudo-stipules present).

Leaves usually alternate, simple or compound, often paripinnate, rachis usually ending in a free tip; leaf or leaflets margins entire to serrate; climbers often have forked tendrils, often on the inflorescences.

Flowers bisexual, sepals 4 or 5, petals 0 or 2–6; disk present; stamens free, in 2 rows of 4 or 5; filaments often hairy.

Ovary superior, usually 3-lobed, 1 ovule per cell.

Fruit capsules, drupes or samaras; capsules usually splitting loculicidally; seeds shiny, aril present.

Confused with:

Anacardiaceae
Sap present, white turning black on exposure.

Burseraceae
Resiniferous; leaves imparipinnate, often with pseudo-stipules; leaflets opposite and with distinct petiolules, with strong odour when crushed, 2 ovules per cell.

Meliaceae
Leaves rarely ending in a free rachis tip; stamens fused in a tube; 1–many ovules per cell; fruit usually globose.

Free rachis tip of *Harpullia arborea*

Fruit and leaves of *Harpullia arborea*

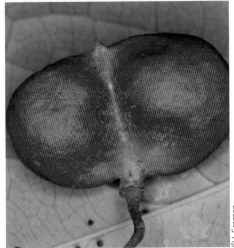

Fruit of *Harpullia arborea*

Sapindaceae cont.

Flower with hairy stamens

Genera:

Allophylus	Trees, shrubs or climbers; leaves palmate compound; petals 4; fruit a globose drupe.
Cardiospermum	Herbaceous or woody climber; leaves bipinnately compound; inflorescences with tendrils; fruit capsular, 3-lobed.
Dimocarpus	Trees; leaves pinnately compound, often with stellate hairs, glands on the lower surface; fruit drupe-like, globose, 1-lobed, warty to spiny.
Dodonaea	Trees; leaves simple, scaly hairs present; young parts viscid; fruit a 3-winged capsule.
Guioa	Trees or shrubs; leaves paripinnate compound; sepals free and unequal; fruit a loculicidal, 3-lobed capsule.
Harpullia	Tree or shrubs; leaves paripinnate compound, with hairs in tufts; flowers with free sepals and 5 petals; fruit a loculicidal, 2–3-lobed capsule.
Lepisanthes	Trees; leaves pinnately compound, lower leaflets reflexing into pseudo-stipules; sepals free and unequal; petals with a scale; fruit a drupe, smooth.
Mischocarpus	Trees or shrubs; leaves paripinnate compound; petals 0–3; fruit a loculicidal capsule.
Nephelium	Trees to shrubs; leaves paripinnate compound; fruit sub-globose, 1–2-lobed, warty to spiny, drupe-like.
Paranephelium	Trees or shrubs; leaves imparipinnate compound; fruit a loculicidal, warty to spiny capsule.
Pometia	Trees; leaves pinnately compound, pseudo-stipules present; fruit drupe-like, globose, smooth.
Xerospermum	Trees; leaves pinnately compound; fruit drupe-like, globose, warty to spiny.

Sapotaceae

Field characters:

Trees or shrubs, white sap present; leaves usually alternate, simple, margins entire, T-shaped hairs present; flowers in simple fascicles, sepals free; ovary superior; seeds shiny, with a large scar along its length.

Description:

Habit trees or shrubs.

Sap white, present at least in the twigs, usually also in the bark.

Stipules present or absent, often caducous.

Leaves alternate or spirally arranged, rarely opposite, often crowded into false whorls, simple, margins entire, often coriaceous; hairs T-shaped, often reddish-brown on the lower surface.

Flowers bisexual, small, regular, often arranged in fascicles (a cluster of flowers all arising from the same point), sometimes solitary, often axillary or from nodes on old wood; sepals free or basally connate, in two whorls of 2, 3 or 4 or in a single whorl of 5; petals often equal in number to sepals, in a single whorl fused at the base, margins sometimes fringed; stamens equal in number to and opposite petals, or more numerous and inserted at base or at throat of corolla tube.

Ovary superior; style simple, often persistent.

Fruit fleshy, usually a berry, indehiscent; seeds 1 or a few, with a shiny bony testa and a conspicuous scar where attached to the fruit wall.

Confused with:

Ebenaceae

Sap absent; leaves with glands on the lower surface; stipules absent; flowers unisexual.

Myrsinaceae

Sap absent; leaves with dark dots; stipules absent; flowers in branched inflorescences.

Reddish-brown lower leaf surface

© R. de Kok

Fruit and seed with scar

© T. Utteridge

Twig with sap

© R. de Kok

Habit and flower of *Payena* sp.

© M. Coode

Genera:

Chrysophyllum	Leaves alternate or spirally arranged; stipules absent; sepals in a single whorl of 4–6, free.
Madhuca	Leaves alternate or spirally arranged; stipules present; sepals in 2 whorls of 2, free; corolla usually hairy at the throat.
Palaquium	Leaves spirally arranged; small stipules usually present; sepals in 2 whorls of 3, free or slightly connate; corolla usually glabrous.
Payena	Leaves alternate or spirally arranged; stipules present; sepals in 2 whorls of 2, fused at base; corolla usually glabrous.
Pouteria	Leaves spirally arranged, rarely sub-opposite; stipules absent; sepals in a single whorl of 4–6, free; corolla sometimes fringed-ciliate or papillose.
Sarcosperma	Leaves opposite to sub-opposite; stipules present; sepals in a single whorl of 4–5, free.

Saxifragaceae

Field characters:

Trees or shrubs, rarely lianas; leaves usually simple, margins serrate; stipules sometimes present; flowers regular, disk present; ovary often inferior, styles 1–5; fruit usually with many seeds.

Description:

Habit trees or shrubs, rarely climbers or herbs.

Sap absent.

Stipules sometimes present.

Leaves opposite to alternate, simple or compound, margins entire to serrate; stellate hairs sometimes present.

Flowers bisexual, regular, 4–7-merous, sometimes petals absent, disk present; stamens as many as sepals to many.

Ovary superior to inferior, 1–5 styles.

Fruit a capsule or follicle, usually with many seeds.

Confused with:

Celastraceae
Stamens same in number or fewer than the sepals, opposite the sepals; ovary superior.

Cornaceae
Stipules absent; leaves entire; fruit a drupe, often blue; ovary inferior.

Cunoniaceae
Leaves compound, opposite; interpetiolar stipules present; ovary superior to semi-inferior.

Rosaceae
Stipules usually present; ovary superior.

Flower and fruit of *Polyosma* sp.

© M. Coode

Fruit of *Polyosma cestroides*

© T. Utteridge

Genera:

Deutzia	Shrubs; leaves opposite, simple, with stellate hairs and scales; stipules absent; ovary (semi-)inferior; fruit a capsule.
Dichroa	Herbs; leaves opposite, simple; stipules absent; ovary semi-inferior; fruit a blue berry.
Hydrangea	Shrubs or climbers; leaves opposite, simple, stellate hairs present; stipules absent; ovary (semi-)inferior; fruit a capsule.
Itea	Trees and shrubs; leaves alternate, simple; stipules present; ovary superior to semi-inferior, with a long forked style; fruit a capsule.
Polyosma	Trees or shrubs; leaves opposite to sub-opposite, simple; stipules absent; ovary superior to inferior, 1 long style; fruit a 1-seeded berry.
Quintinia	Shrubs or trees, rarely climbers; leaves alternate, simple, with peltate scales; stipules absent; ovary inferior, fruit a capsule.

Scrophulariaceae

Field characters:

Herbs, sometimes shrubs; leaves opposite or alternate, simple, usually entire; inflorescence (in)determinate.

Description:

Habit herbs, sometimes shrubs or small trees, some species are hemiparasitic herbs; stems sometimes square.

Sap absent.

Stipules absent.

Leaves opposite or alternate, simple, margins serrate to entire, sometimes with anisophyllous leaves; some species turning dark blue-black when dried.

Inflorescence (in)determinate.

Flower bisexual, zygomorphic, fused corollas, usually 5 corolla lobes; stamens (2–)4 fused to the corolla lobes.

Ovary superior, 2 carpels; style terminal and persistent.

Fruit a capsule with many seeds; seeds sometimes winged

Confused with:

Acanthaceae
Stems with nodes swollen when fresh or shrunken when dried; leaves sometimes with cystoliths; fruit sometimes with hooks inside; seeds few.

Gesneriaceae
Leaves often hairy and fleshy; fruit 1-locular.

Lamiaceae
Fruit few seeded, style not persistent in fruit.

Verbenaceae
Flowers trumpet-shaped; fruit few seeded.

Flower of *Brookea* sp.

Habit of *Torenia* sp.

Scrophulariaceae cont.

© R. de Kok

Habit of *Limnophila* sp.

Genera:	
Angelonia	Herbs or shrubs, erect or creeping; leaves opposite to spirally arranged; fertile stamens 4; fruit a loculicidally 2-valved capsule; seeds not winged.
Brookea	Small trees or shrubs; leaves opposite, densely hairy; fertile stamens 4; fruit a capsule, seeds not winged.
Buchnera	Hemiparasites, plants turning dark blue-black when dried; leaves opposite to spirally arranged; corolla tube straight, fertile stamens 4; fruit a loculicidally 2-valved capsule; seeds not winged.
Cyrtandromoea	Shrubs; leaves opposite, anisophyllous; fertile stamens 4; fruit a dehiscent capsule; seeds not winged.
Limnophila	Herbs, often growing in or near water, stem with air cavities, aromatic when bruised; leaves opposite or in whorls; fertile stamens 4; fruit a septicidally 4-valved capsule; seeds ribbed.
Lindernia	Herbs with quadrangular stem; leaves opposite; fertile stamens 2 or 4; fruit a septicidally 2–3-valved capsule; seeds not winged.
Striga	Hemiparasites, plants turning dark blue-black when dried; leaves opposite to spirally arranged; corolla tube curved, fertile stamens 4; fruit a loculicidally 2-valved capsule; seeds not winged.
Torenia	Herbs with quadrangular stem; leaves opposite; calyx winged; fertile stamens 4; fruit a septicidally 2-valved capsule; seeds not winged.
Wightia	Trees or stranglers; leaves opposite; fertile stamens 4; fruit a septicidally 2-valved capsule; seeds winged.

Simaroubaceae

Field characters:

Trees and shrubs; leaves spirally arranged; flowers petals free; ovary superior, disk present.

Description:

Habit trees and shrubs; bark usually with a bitter taste.

Sap absent.

Stipules absent or present, leaving a circular scar when falling off.

Leaves spirally arranged, simple or pinnate, margins usually toothed, sometimes entire, often with pitted or flattish glands on the lower surface.

Flowers bisexual, small, regular, 3–5-merous, petals free.

Ovary superior, disk present, often deeply lobed to apocarpous, 4–5-locular, 1 ovule per cell.

Fruit usually indehiscent, often drupaceous, sometimes a samara.

Confused with:

Burseraceae
Leaves compound with opposite leaflets and swollen petiolules; resins present.

Meliaceae
Stamens usually united into a tube; ovary not deeply lobed.

Rutaceae
Leaves with pellucid dots; bruised leaves with an orange/citrus smell.

Sapotaceae
White sap present; T-hairs often present on the under-surface of the leaves.

© T. Utteridge

Eurycoma sp.

Simaroubaceae cont.

Flower of *Quassia indica*

Genera:	
Ailanthus	Trees; leaves pinnate; margins entire or toothed; large glands at base of leaflets; fruit a samara.
Brucea	Tree or shrubs; leaves pinnate, margins serrate, rarely entire; stamens same in number as petals; fruit a drupe.
Eurycoma	Small trees; leaves pinnate, with swollen rachis nodes; leaflets when gently broken are still attached with thin white treads; fruit a nut.
Harrisonia	Scrambler with stipular thorns; leaves pinnate or palmate, petiole winged; fruit a drupe.
Irvingia	Trees; leaves simple; stipules present, which leave an annular scar; fruit a drupe.
Picrasma	Trees or shrubs; leaves pinnate, margins entire; stipules present; fruit drupe-like.
Quassia	Trees or shrubs; leaves simple or pinnate, rachis sometimes winged, with glands on under-surface; carpels free; fruit drupaceous or woody, often compressed laterally.
Soulamea	Trees and shrubs; leaves simple; fruit dry, indehiscent, flattened, with wings.

Solanaceae

Field characters:

Leaves alternate, simple, sometimes with prickles; stipules absent; corolla lobes fused to various extents, ovary superior.

Description:

Habit trees, shrubs, lianas and herbs.

Sap absent.

Stipules absent.

Leaves alternate, simple, margins entire but often lobed or deeply divided, penninerved; sometimes with prickles.

Flowers bisexual, usually regular, (4–)5-merous, calyx fused; corolla usually contorted in bud, fused, often showy; anthers with pores.

Ovary superior, often 2-celled.

Fruit usually a berry; seeds many.

Confused with:

Boraginaceae
Inflorescence a 1-sided cyme; ovules 1 per cell; style mostly gynobasic; fruit not a berry.

Convolvulaceae
Twining climbers; calyx lobes not fused and overlapping each other; corolla folded.

© R. de Kok

Dried specimen of *Lycianthes biflora* with inflorescences in leaf axis

© R. de Kok

Dried specimen of *Solanum torvum* with extra-axillary inflorescences

Genera:

Lycianthes	Shrub or herb; inflorescences always in a leaf axis; calyx 10-lobed.
Physalis	Herbs; inflorescences terminal or seemingly axillary; calyx 5-lobed, accrescent when fruiting and covering the fruit.
Solanum	Small tree, shrub or herb; inflorescences terminal or extra-axillary (inflorescence arising from the shoot but not associated with a leaf); calyx 5-lobed.

Sparrmanniaceae (Tiliaceae)

Field characters:

Shrubs, small trees or herbs, often with stellate hairs; leaves alternate or spirally arranged, simple, often 3-nerved from the base; flowers bisexual, 5-merous, regular; stamens usually free; fruit often with spines.

Description:

Habit usually shrubs, rarely scandent or small trees or herbs, often with stellate hairs; bark fibrous.

Sap absent.

Stipules present, but often falling off early.

Leaves alternate or spirally arranged, simple or rarely lobed, margins usually toothed, often 3-nerved from the base or palminerved, often with domatia.

Flowers bisexual, 5-merous, showy; calyx lobes free, often with an epicalyx; petals are generally clawed; stamens usually more than (4–)15, in multiple whorls, free or rarely united in bundles or a single central bundle, inserted opposite the petals.

Ovary superior.

Fruit about half of the genera have fleshy or fibrous indehiscent fruit whereas the remainder are dehiscent, then often spiny; seeds numerous, small, sometimes winged.

Confused with:

Brownlowiaceae
Stamens 5–20 in a single whorl.

Byttneriaceae
Stamens usually 5(–10), free to the base and sometimes clustered in bundles, inserted opposite the sepals.

Durionaceae (Bombacaceae)
Leaves simple to compound, margins entire, covered with scales; stamens often connate into bundles; seeds with an aril.

Malvaceae
Epicalyx usually present; stamens fused in a tube and attached to the corolla.

Sterculiaceae
Flowers usually with a single perianth whorl; stamens usually 5; carpels divided.

Three-veins from the base of the leaf of *Grewia* sp.

Flower of *Grewia* sp.

© R. de Kok

Flower of *Microcos* sp.

Major genera:	
Berrya	Small trees; leaves toothed at apex only; fruit a capsule with winged seeds.
Colona	Trees; leaves toothed along the whole of their margin; stipules leaf-like; fruit a 3–4-winged achene.
Grewia	Shrub or climbers; leaves toothed along the whole of the margins or at apex only; fruit a drupe.
Microcos	Tree or shrub; leaves toothed at apex only, sometimes with scales; fruit a drupe.
Pentace	Trees; leaves toothed at apex only, sometimes with scales, petioles swollen apically; fruit a samara.
Schoutenia	Trees; leaves toothed at apex only, sometimes with scales, base unequal; fruit a capsule with a persistent enlarged papery calyx.
Trichospermum	Trees; leaves toothed along the whole of the margin; glands at base; petioles swollen apically; fruit a flattened 2-valved capsule, seeds with long white hairs.
Triumfetta	Herb; leaves margins serrate; stamens 4–5; fruit a bristly capsule.

Sterculiaceae

Field characters:

Usually trees or shrubs; leaves usually simple, alternate, often with stellate hairs or scales; petiole often thickened at base and apex; flowers 5-merous, actinomorphic; stamens free or connate; carpels divided.

Description:

Habit trees or shrubs with a particularly fibrous bark.

Sap absent but producing mucilage when wounded .

Stipules usually inconspicuous and caducous.

Leaves usually spirally arranged; simple, some are palmately lobed (fewer palmately compound); margins entire; palmi- to pinnately-nerved; stellate hairs are characteristic but are sometimes only found on shoot apices or flowers.

Flowers bisexual or unisexual, 5(–3)-merous, sepals 3–5, partly connate, rarely free, often with epicalyxes; petals 5 or absent; stamens free or fused into a terminated head or bundle.

Ovary superior, carpels free.

Fruit carpels clearly separated, with each carpel usually developing into a fruitlet or mericarp, seed aril absent.

Flowers of *Sterculia* sp.

© T. Utteridge

Confused with:

Byttneriaceae

Shrubs, rarely trees, lianas or herbs; stamens 5(–15), in a single whorl.

Helicteraceae

Trees or shrubs; flowers zygomorphic, 2 whorls of perianth.

Durionaceae (Bombacaceae)

Trees, leaves often with scales on the lower surface; epicalyx absent; 2 whorls of perianth; stamens numerous, seeds often glabrous.

Malvaceae

Shrubs or small trees; epicalyx usually present; 2 whorls of perianth; stamens many, united in a long tube, at base attached to the corolla.

Sparrmanniaceae (Tiliaceae)

Shrubs or small trees; leaves 3-veined from the base, often with domatia; 2 whorls of perianth; stamens many, free or rarely united in bundle(s), then not fused to the corolla.

Fruit of *Sterculia stipulata*

© J. Gregson

© RBG Kew

Commersonia bartramia

Major genera:

Abroma	Shrubs; leaves simple, sometimes lobed, margins serrate to entire; flowers with sepal and petals; stamens 10; fruit a capsule with 5 wings.
Commersonia	Small trees or shrubs; leaves simple, margins serrated; flowers with sepals and petals; stamens 5; fruit globose capsule, covered stiff hairs.
Firmiana	Trees; leaves simple and tri-lobed, margins entire; flowers with only petals; stamens 15–25; fruit a winged capsule.
Heritiera	Trees; leaves simple to palmately compound, margins entire; flowers with only petals; stamens 8–10; fruit a samara.
Kleinhovia	Small tree; leaves simple, margins entire; flowers with sepal and petals; stamens 10; fruit an inflated thin-walled capsule.
Pterocymbium	Trees; leaves simple, margins entire; flowers with only petals; stamens <10; fruit 5 follicles with a long broad-shaped wing.
Pterospermum	Trees; leaves simple, margins entire to serrate; flowers with sepals and petals; stamens 12–15; fruit a woody capsule, smooth or angled or winged.
Scaphium	Trees; leaves simple, margins entire; flowers with only petals; stamens 8–10; fruit a follicle with a long broad wing.
Sterculia	Trees to shrubs, leaves simple to palmately compound, margins entire; flowers with only petals; stamens 10–15; fruit coriaceous or woody, not winged, smooth.

Symplocaceae

Field characters:

Small trees; leaves simple, penninerved; stipules absent; flowers actinomorphic; stamens numerous; ovary inferior; fruit 2–5-celled drupes.

Description:

Habit small trees or shrubs.

Sap absent.

Stipules absent.

Leaves alternate or spirally arranged, simple, margins usually dentate, penninerved, often yellowish when dried.

Flowers usually bisexual, actinomorphic, petals and sepals fused at least at base, usually each flower subtended by 1 bract and 2 bracteoles; stamens usually many.

Ovary inferior, ovules pendulous 2–4 per cell.

Fruit 2–5-celled drupe.

Confused with:

Alangiaceae
Leaf base unequal; petals free; stamens not numerous; ovule 1 per cell; fruit a berry.

Ebenaceae
Leaves entire with glands on the underside; ovary superior; fruit a berry.

Rosaceae
Stipules present.

Theaceae
Ovary superior; fruit a berry or capsule.

Symplocos sp.

© M. Coode

Symplocos pendula

© G. Lewis

Symplocos sp.

© S. Andrews

Genus:

Symplocos	As for the family.

Theaceae

Field characters:

Trees and shrubs; leaves alternate or spirally arranged, simple, penninerved; stipules absent; sepals and petals free, imbricate.

Description:

Habit trees and shrubs.

Sap absent.

Stipules absent.

Leaves alternate or spirally arranged, simple, margins entire to usually dentate, penninerved.

Inflorescences flowers usually solitary.

Flowers bisexual, usually big, usually 5-merous; sepals and petals free, imbricate; stamens usually numerous, often connate with petals at base; sometimes with paired bracteoles.

Ovary usually superior.

Fruit dehiscent leathery capsule or berry.

Confused with:

Guttiferae
Usually with yellow sap; leaves opposite; stigma large, flat.

Symplocaceae
Ovary inferior; fruit a drupe.

Flower of *Tetramerista* sp.

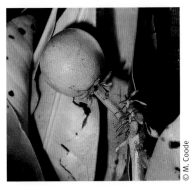

Fruit of *Tetramerista* sp.

Flower of *Ternstroemia* sp.

Theaceae cont.

Flower of *Gordonia* sp.

Genera:	
Adinandra	Trees and shrubs; flowers axillary, solitary or several together; 1 undivided style; fruit indehiscent, many-seeded berry.
Camellia	Trees or shrubs; flowers axillary, solitary or several together; styles 3–5 or 1 divided into 3–5 arms; fruit a woody capsule that splits open at apex, seeds not winged.
Eurya	Trees and shrubs; flowers axillary fascicles or solitary; 1 undivided style; fruit a fleshy berry.
Gordonia	Trees; flowers axillary, solitary; 1 undivided style; fruit a woody capsule that splits open at apex; seeds winged.
Pyrenaria	Trees and shrubs; flowers axillary, solitary; style 5-branched; fruit drupaceous; seeds not winged.
Schima	Trees; flowers axillary or sub-terminal, solitary or several together; 1 undivided style; fruit woody capsule that splits open at apex; seeds winged.
Ternstroemia	Trees and shrubs, sap brown; flowers usually extra-axillary, solitary; 1 undivided style; fruit a fleshy or corky berry; seeds 1–4, flattened.
Tetramerista	Trees; flowers in an axillary panicle; fruit a berry.

Thymelaeaceae

Field characters:

Bark often fibrous and difficult to break; leaves opposite to alternate, simple, margins entire; stipules absent; flowers bisexual, regular; calyx united at base at least; corolla absent or reduced to small lobes; ovary superior.

Description:

Habit trees, shrubs, climbers, rarely herbs; bark silky fibrous, tough.

Sap absent.

Stipules absent.

Leaves opposite to alternate, simple, margins entire, sometimes with translucent dots.

Flowers bisexual, 5–6-merous, regular; calyx forming a 'longish' tube; corolla absent or present as small lobes sitting on the inside of the calyx tube; stamens 2–many in 1–3 whorls, usually twice as many as the calyx lobes.

Ovary superior.

Fruit a drupe, berry or capsule; seeds 1–3, with an appendage or aril.

Confused with:

Chrysobalanaceae
Stipules present; flower zygomorphic, calyx and corolla present.

Icacinaceae
Bark without silky fibres; corolla usually present.

Oleaceae
Leaves opposite; corolla present; stamens 2.

Rosaceae
Stipules usually present; calyx and corolla well developed.

© J. Dransfield

Flower of *Wikstroemia* sp.

© M. Coode

Flowers of *Gonystylus* sp.

Thymelaeaceae cont.

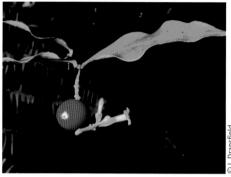

Fruit of *Wikstroemia* sp.　　　　　　Flowers of *Linostoma* sp.

Genera:	
Amyxa	Trees; leaves alternate, translucent dots present, few lateral veins; stamens free; fruit a capsule.
Aquilaria	Trees or shrubs; leaves alternate, without translucent dots; lateral veins parallel; fruit a capsule.
Enkleia	Climbers often with hooks; leaves alternate, without translucent dots, lateral veins reticulate; fruit a drupe.
Gonystylus	Trees; leaves alternate, translucent dots, with many lateral veins; stamens free; fruit a capsule.
Linostoma	Shrubs or climbers; leaves opposite, without translucent dots, lateral veins parallel; fruit a drupe.
Phaleria	Shrubs or trees; leaves opposite, without translucent dots, blade >11 cm long, lateral veins reticulate; ovary 2-locular; fruit a drupe.
Pimelea	Herb; leaves opposite to alternate; fruit a drupe.
Wikstroemia	Shrubs or trees; leaves opposite, without translucent dots, blade <13 cm long, lateral veins reticulate; ovary 1-locular; fruit a drupe.

Ulmaceae

Contributed by Melanie Thomas

Field characters:

Trees and shrubs; leaves alternate, simple, often unequal at the base, often 3-veined; flowers unisexual; stamens opposite the single perianth whorl; ovary superior, stigmas paired.

Description:

Habit trees or shrubs.

Sap absent.

Stipules present, sometimes overlapping each other, falling off early, usually membranous.

Leaves alternate, simple, often with unequal sides at the base; margins serrate to rarely entire; pinnately veined or 3-veined from the base.

Flowers usually unisexual, regular, (4–)5(–7)-merous; solitary or in lax simple inflorescences; perianth a single whorl; tepals large, conspicuous in the male flowers with the stamens opposite them; tepals often not conspicuous in female flowers.

Ovary superior, carpels 2; style 1, dividing into conspicuous paired diverging stigmas.

Fruit a drupe.

Confused with:

Cannabaceae
Herbs; leaves compound.

Moraceae
White sap present; stipules conspicuous; flowers often asymmetrical, often crowded into dense heads or inside figs.

Tiliaceae
Stellate hairs present; bisexual flowers with petals; anthers numerous; stigma 1.

Urticaceae
Small trees to herbs; leaves with cystoliths; stipules usually conspicuous, persistent; flowers crowded into dense heads; stigma 1.

Rhamnaceae
Trees, shrubs or lianas often spiny; flowers bisexual; petals present, conspicuous disc present.

© R. de Kok

Fruit of *Trema* sp.

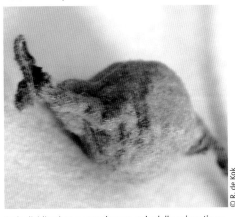

© R. de Kok

Style dividing into a conspicuous, paired diverging stigma

Ulmaceae cont.

Flower of *Trema* sp.

Genera:	
Aphananthe	Leaves leathery, 3-veined at base or pinnately veined, margins entire or almost so; stipule scar not circular; flowers often solitary in axils and long-pedicellate, male without pistillode, female without staminodes; fruit round in cross-section.
Celtis	Leaves often leathery, 3-veined from base, margins often entire; stipule scar not circular; inflorescences few-flowered, male with pistillode, female with staminodes; fruit round in cross-section with long pedicel.
Gironniera	Leaves leathery, pinnately many-veined, margins entire or almost so; stipules scar circular; inflorescences often few-flowered; fruit laterally compressed, lens-shaped in cross-section, sub-sessile or short-pedicellate.
Trema	Leaves thin-textured, 3-veined at base or pinnately veined, margins conspicuously serrate; stipule scar not circular; inflorescence many-flowered; fruit laterally compressed, lens-shaped in cross-section, sub-sessile or short-pedicellate.

Urticaceae

Contributed by Melanie Thomas

Field characters:

Herbs to trees; leaves simple, cystoliths present, margins often serrate, white beneath, often 3-veined from the base; flowers tiny, unisexual, often grouped into compact inflorescences; perianth a single whorl or absent; stigma 1.

Description:

Habit herbs, shrubs or trees; sometimes with stinging hairs.

Sap absent.

Stipules usually present, often conspicuous, usually ± fused, intrapetiolar, very rarely interpetiolar, sometimes lateral and free.

Leaves opposite or alternate, simple, margins entire to serrated, often asymmetrical, often discolorous because of white tomentum beneath; veins often conspicuous, often 3-veined from the base, otherwise pinnately veined; cystoliths conspicuous in dried specimens, either dot-like or elongate.

Flowers unisexual, tiny, green; inflorescences can consist of only 1 sex or both; usually grouped into axillary clusters and branched, spike-like, globular often fleshy heads, or broad flat receptacles; perianth a single whorl or absent; stamens inflexed, opposite the tepals.

Ovary superior, stigma 1.

Fruit achenes, tiny, sometimes with winged or fleshy perianth.

Confused with:

Cannabaceae
Herbs; leaves palmate-compound, not discolorous.

Moraceae
White sap present; leaves concolorous, cystoliths absent; stamens rarely inflexed; stigmas often 2.

Ulmaceae
Trees and shrubs; flowers never aggregated into a dense inflorescence-head; stamens not inflexed; stigmas 2.

Euphorbiaceae
White sap sometimes present; cystoliths absent; stigmas 3; fruit usually dehiscing into 3 parts leaving a central columella.

© J. Gregson

Dendrocnide sp.

© T. Utteridge

Fruit of *Oreocnide obovata*

Urticaceae cont.

© A. Moore

Debregeasia squamata

© RBG Kew

Habit of *Pipturus* cf. *asper*

© RBG Kew

Flowers of *Poikilospermum suaveolens*

Urticaceae cont.

Major genera:	
Boehmeria	Herbs to trees; leaves often opposite, often 3-veined from the base, margins usually toothed; inflorescence mostly single or several axillary spikes, sometimes in axillary clusters or branched; perianth dry, often ribbed or winged; stigma conspicuous, thread-like, persistent.
Cypholophus	Herbs to trees; leaves opposite; often 3-veined from the base, margins usually toothed; inflorescence single, axillary, sessile, dense globose heads; perianth fleshy; stigma minute, curled.
Debregeasia	Shrub or small tree; leaves alternate, 3-veined from the base, often discolorous, margins toothed; inflorescences dichotomously branched with orange fleshy clusters; stigma capitate.
Dendrocnide	Trees or shrubs, stinging hairs present; leaves spirally arranged, leathery with prominent veins, sometimes 3-veined, margins mostly entire; inflorescence lax-branched, often long and narrow; stigma filiform; fruit often flattened, asymmetrical.
Elatostema	Herbs, rarely shrubs, with elongate cystoliths; leaves alternate, often distichous, margins entire to toothed; inflorescence lax-branched or a flat receptacle; stigma capitate; fruit often aggregated into heads.
Laportea	Herbs with stinging hairs; leaves alternate, thin-textured, margins toothed; inflorescence usually lax-branched; stigma thread-like; fruit much laterally compressed, asymmetrical with excentric thread-like stigma.
Leucosyke	Shrubs; leaves alternate, 3-veined from the base, discolorous; inflorescences paired, axillary, short-stalked, tight globose heads; stigma capitate; fruit often aggregated into paired, stalked, globose clusters.
Maoutia	Shrubs; leaves alternate, 3-veined from the base, discolorous; inflorescences loose, dichotomously branched with many tiny clusters; stigma capitate; fruit slightly flattened.
Nothocnide	Herbaceous climber; leaves often opposite, often 3-veined from the base, margins entire, concolorous; inflorescences mostly single or several axillary spikes, sometimes axillary clusters or branched; perianth dry, often ribbed or winged; stigma conspicuous, thread-like, persistent.
Oreocnide	Herbs to trees; leaves alternate, often 3-veined from the base, margins usually almost entire; inflorescences sessile clusters; perianth dry, often ribbed or winged; stigma capitate, fruit surrounded by white fleshy cup-like receptacle.
Pilea	Herbs with elongate cystoliths; leaves opposite, often fleshy and drying black; inflorescence lax-branched; stigma capitate.
Pipturus	Herbs to trees; leaves alternate, often 3-veined from the base, margins toothed, usually discolorous; inflorescences mostly single or several axillary spikes, sometimes axillary clusters or branched; perianth succulent; stigma conspicuous, thread-like, persistent; fruit apex with conspicuous dark aperture after stigma is shed.
Poikilospermum	Woody climbers; leaves spirally arranged, leathery, veins prominent, sometimes 3-veined from base, margins entire; inflorescences paired or several times dichotomous, mostly with tight purple globose heads; bracts large, persistent; stigma capitate.
Pouzolzia	Herbs to trees; leaves often opposite, often 3-veined from the base, margins mostly entire; inflorescences mostly loose axillary clusters; perianth dry, often ribbed or winged; perianth dry, often ribbed or winged; stigma caducous; fruit without a dark aperture at apex.

Verbenaceae

Field characters:

Leaves opposite; stipules absent; flowers trumpet-shaped (salverform); inflorescences indeterminate; fruit varied; usually growing in disturbed areas.

Description:

Habit herbs, shrubs, small trees or lianas.
Sap absent.
Stipules absent.
Leaves opposite, decussate, simple, margins entire to serrate.
Inflorescences indeterminate, terminal or axillary, racemose.
Flowers bisexual, zygomorphic, salverform, occasionally somewhat bilabiate; stamens 4, not or only just exceeding the corolla tube.
Ovary superior, style terminal.
Fruit a schizocarp, (2–)4 nutlets, or a drupe with 2–4 stones.

Confused with:

Lamiaceae
Inflorescence determinate; flowers usually 2-lipped; stamens exceeding the corolla tube.

Acanthaceae
Leaves with cystoliths usually present; stamens usually exceeding the corolla tube; fruit a splitting capsule, often with hooks on the inner surface.

Flowers of *Citharexylum caudatum*

Habit of *Stachytarpheta* sp.

Fruit of *Citharexylum caudatum*

Flowers of *Lantana camara*

Flowers of *Duranta* sp.

Genera:	
Citharexylum	Tree; fruit with two 2-seeded pyrenes; style 2-lobed.
Duranta	Shrubs, sometimes with spines; style club-shaped; fruit with four 2-seeded pyrenes.
Lantana	Shrubs, often with spines; drupes fleshy.
Lippia	Herbs or shrubs; leaves with simple hairs; fruit dry.
Phyla	Herbs, creeping; leaves with T-hairs.
Petrea	Lianas; leaves feel rough when touched; calyx lobes enlarged.
Stachytarpheta	Herbs; flowers with 2 stamens.
Verbena	Herbs; fruit with 4 single-seeded pyrenes

Violaceae

Field characters:

Trees and shrubs; nodes irregularly spaced; leaves spirally arranged, simple; stipules present; flowers 5-merous; ovary superior; fruit a capsule.

Description:

Habit trees and shrubs.

Sap absent.

Stipules present; striate on surface, especially when dry.

Leaves spirally arranged, simple, margins usually dentate; venation often scalariform; drying pale yellow-green.

Flowers usually bisexual, 5-merous, usually zygomorphic; sepals and petals usually free; filament usually fused and connectives with appendages.

Ovary superior, 1- or 3-locular, placentation parietal.

Fruit a capsule, dehiscent usually into 3 parts, lacking a central columella; seed numerous.

Confused with:

Euphorbiaceae
Flowers unisexual; placentation never parietal; fruit with a central columella.

Flacourtiaceae
Stamens numerous; fruit rarely a capsule.

Theaceae
Flowers large and usually solitary; stipules absent.

© M. Coode

Fruit of *Rinorea* sp.

© R. de Kok

Striped stipules of *Rinorea* sp.

Genera:

Hybanthus	Shrubs; stipules small, not striate; flowers zygomorphic.
Rinorea	Shrubs and trees; stipules striate lengthwise; flowers regular.

Vitaceae

Field characters:

Lianas or climbing herbs, climbing by leaf-opposed tendrils; inflorescences cymose, leaf-opposed; flowers small; ovary superior; fruit fleshy berries.

Description:

Habit lianas or climbing herbs.

Sap absent.

Stipules present, sheathing stipules along petiole margin.

Leaves spirally arranged, simple to palmately compound, palmately veined, margins serrate; tendrils leaf-opposed, rarely lacking.

Inflorescences often opposite the leaves, sometimes with tendrils, cymose, very rarely flattened or ribbon-like.

Flowers bisexual, rarely unisexual, 4–5-merous; petals valvate; stamens free, equal in number to and opposite the petals.

Ovary superior.

Fruit a fleshy berry; seeds 1–4.

Confused with:

Convolvulaceae
Climbers, rarely small trees, tendrils absent; flowers bisexual.

Cucurbitaceae
Tendrils at 90° relative to a leaf; flowers large; ovary inferior; fruit many-seeded.

Leeaceae
Climbers to small trees, tendrils absent; stamens fused in a tube; ovary superior.

Passifloraceae
Climbers, tendrils in leaf axils; stipules present; glands on petiole or leaf base; fruit many-seeded.

Cissus simplex

Fruit of *Pterisanthes* sp.

Vitaceae cont.

© T. Utteridge

Tetrastigma sp.

Major genera::	
Ampelocissus	Shrubs or climbing herbs; leaves simple to palmately compound, glaucous below; inflorescences leaf-opposed, bearing tendrils, axis not flattened when fruiting; flowers bisexual, 4–5-merous; berry with 2–4-seeds.
Cayratia	Climbing shrubs or creeping herbs; leaves palmately compound; inflorescences leaf-opposed or seemingly terminal without tendrils, axis not flattened when fruiting; flowers bisexual, 4-merous; berry 2–4-seeded.
Cissus	Climbing shrubs; leaves simple; inflorescences leaf-opposed without tendrils, axis not flattened when fruiting; flowers bisexual, 4-merous; berry single-seeded.
Pterisanthes	Climbing shrubs; leaves simple to palmately compound; inflorescences leaf-opposed, bearing tendrils, axis flattened when fruiting; flowers bisexual, 4–5-merous; berry with 1–3 seeds.
Tetrastigma	Lianas or climbing shrubs; leaves palmately compound; inflorescences axillary or leaf-opposed without tendrils, axis not flattened when fruiting; flowers unisexual, 4-merous; stigma 4-lobed; berry with 1–4 seeds.

Winteraceae

Field characters:

Shrubs or trees, glabrous; leaves spirally arranged, simple, margins entire, penninerved; stipules absent; ovary superior.

Description:

Habit shrubs or trees, rarely epiphytes.
Sap absent.
Stipules absent.
Leaves spirally arranged, simple, entire, often with pellucid dots, penninerved, glabrous, often glaucous and sharp-tasting.
Flowers usually unisexual, usually regular; sepals 2–6, valvate to spirally inserted, free or fused; petals absent to many, free, in whorls or spirals; stamens 5 to many.

Ovary superior, carpels often free.
Fruit berry-like or a cluster of follicles developed from free carpels.

Confused with:

Illiciaceae
Leaves crowded at the apex of twigs; tepals not differentiated into petals and sepals; carpels forming a star-shaped fruit.

Magnoliaceae
Trees; stipules present, leaving a circular scar.

Myrsinaceae
Black dots or lines on the leaves, calyx, petals and fruit; petals fused, 5, contorted; fruit a drupe from a single carpel.

© M. Coode

Drimys sp.

Genera:

Drimys	As for the family.

Zingiberaceae

Field characters:

Herbs, rhizomes often aromatic; leaves 2-ranked, with an open sheath bearing an open ligule; tepals fused into a tube, at least at base; stamens only 1 fertile; ovary inferior.

Description:

Habit herbs, often large, rhizomatous, the rhizomes often aromatic.

Sap absent.

Stipules absent, but with a ligule present.

Leaves 2-ranked, with an open sheath, open ligule usually present on leaf sheath; blade usually absent from lower leaves, rolled when young; hairs sometimes branched.

Inflorescences terminal, either on the leafy shoot or on a separate non-leafy shoot (this flowering or fruiting shoot can sometimes be many meters away from the leafy shoots).

Flowers tepals fused into a tube, at least at base; outer whorl (calyx) usually 3-toothed or 3-lobed, inner whorl 3-lobed; stamens only 1 fertile, the others petaloid staminodia.

Ovary inferior, 1- or 3-locular.

Fruit a dry to fleshy, dehiscent or indehiscent capsule; seeds sometimes arillate.

Confused with:

Costaceae
Leaves spirally arranged; with a closed sheath, ligule forming a ring above the pseudo-petiole.

Marantaceae
Rhizomes not aromatic; leaf apex usually offset relative to the midrib; petiole swollen apically; ligule absent.

Orchidaceae
Stamens and style fused together forming a column; roots covered in a papery or velvet sheath.

Flower of *Etlingera* sp.

© R. de Kok

Boesenbergia pulchella

© J. Gregson

Inflorescence of *Plagiostachys* sp.

© J. Gregson

Globba sp.

© T. Utteridge

Major genera:

Alpinia	Robust herbs; inflorescences on a leafy shoot; with lateral staminodes absent or reduced; fruit a capsule.
Amomum	Robust herbs; inflorescences cone-like, without coloured sterile bracts, on a separate non-leafy shoot; lateral staminodes absent or reduced; fruit berry-like or a 3-locular capsule, in dense heads.
Boesenbergia	Small herbs; inflorescence on a leafy shoot; stamens not exceeding lip; lateral staminodes well developed; fruit a 3-locular capsule.
Burbidgea	Robust herbs; inflorescences on a leafy shoot; lateral staminodes absent or reduced.
Etlingera	Robust herbs; inflorescences cone-like, surrounded with sterile coloured bracts, on a separate non-leafy shoot; lateral staminodes absent or reduced; fruit indehiscent, capsular.
Globba	Small herbs; inflorescences on a leafy shoot; flower-lip joined to the stamens; stamens long exserted, bow-like; fruit a capsule, irregularly dehiscing.
Hedychium	Medium-sized herbs; inflorescences on a leafy shoot; flowers with long extended stamens, lateral staminodes well developed; fruit with 3-locular capsule.
Plagiostachys	Robust herbs; inflorescence compact, on a leafy shoot; breaking through the leaves sheaths halfway along the stem; lateral staminodes absent or reduced.
Zingiber	Medium-sized to robust herbs with long creeping rhizomes; inflorescences on a separate non-leafy shoot; flower-lip not divided; anther with a long curved appendage; fruit a 3-locular capsule.

Provisional checklist for Danum Valley (D), Maliau Basin (M) and Imbak Canyon (I)

This list is based on the collections on databases in the Herbarium of the Sabah Forest Department and the Royal Botanic Gardens Kew. Any updates and corrections are welcome.

ACANTHACEAE

Species	M	D	I
Acanthus sp.	M		
Dicleptera aff. bivalvis		D	
Dyschoriste motleyi		D	
Dyschoriste oligosperma		D	I
Eranthemum borneense		D	
Gymnostachyum affine		D	
Gymnostachyum sp.	M		I
Hemigraphis reptans		D	
Hemigraphis sumatrensis	M		I
Hypoestes sp.	M		
Justicia henicophylla	M		I
Lepidagathis sp.	M		
Paragoldfussia sp.		D	
Pseuderanthemum graciliflorum			I
Pseuderanthemum sp.		D	
Ptyssiglottis gibbisae			I
Ptyssiglottis psychotriifolia		D	
Semnostachya galeopsis			I
Staurogyne jaherii	M	D	I
Staurogyne setigera		D	
Strobilanthes sp.		D	

ACTINIDIACEAE

Species	M	D	I
Saurauia acuminata	M		
Saurauia agamae		D	I
Saurauia amplifolia		D	
Saurauia borneensis	M		
Saurauia cf. ferox	M		
Saurauia cf. horrida	M	D	I
Saurauia longistyla		D	I
Saurauia nudiflora		D	
Saurauia palawanensis	M		
Saurauia platyphylla			I
Saurauia reinwardtiana		D	
Saurauia speciosa		D	
Saurauia strigosa	M		
Saurauia subglabra			I
Saurauia sp. (Clark 43)		D	

AGAVACEAE

Species	M	D	I
Pleomele angustifolia	M		

ALANGIACEAE

Species	M	D	I
Alangium ebenaceum	M		
Alangium griffithii		D	
Alangium javanicum	M	D	
Alangium javanicum var. meyeri		D	I
Alangium longiflorum	M		

ANACARDIACEAE

Species	M	D	I
Buchanania arborescens	M	D	
Buchanania insignis		D	
Buchanania sessilifolia	M	D	
Buchanania splendens		D	
Campnosperma auriculatum	M		
Campnosperma squamatum	M		
Gluta aptera	M		
Gluta laxiflora	M		
Gluta sabahana	M		
Gluta speciosa	M		
Gluta wallichii	M		
Koordersiodendron pinnatum	M		
Mangifera bullata	M		
Mangifera foetida	M	D	
Mangifera griffithii	M		
Mangifera indica		D	
Mangifera pajang	M	D	I
Mangifera parvifolia	M		
Mangifera rigida	M		
Mangifera rufocostata		D	
Mangifera swintonioides	M	D	
Melanochyla angustifolia			I
Melanochyla auriculata	M		
Melanochyla bullata	M		
Parishia insignis	M		
Pegia sarmentosa	M	D	
Semecarpus borneensis		D	
Semecarpus bunburyanus		D	
Semecarpus glaucus		D	
Semecarpus lineatus		D	
Semecarpus rufovelutinus			I
Semecarpus trengganuensis		D	
Solenocarpus philippinensis		D	I
Spondias philippinensis		D	

ANISOPHYLLEACEAE

Species	M	D	I
Anisophyllea corneri	M		
Anisophyllea disticha			I
Carrallia bracteata	M		

ANNONACEAE

Species	M	D	I
Anaxagorea borneensis		D	I
Anaxagorea javanica		D	
Artabotrys sp.			I
Artabotrys gracilis		D	
Artabotrys hirtipes		D	
Artabotrys roseus	M	D	
Artabotrys sauveolens	M	D	
Artabotrys sp. A. (Juin 11)		D	
Cyathostemma excelsa	M	D	
Dasymaschalon clusiflorum	M	D	
Desmos dumosus		D	
Disepalum anomalum	M		
Enicosanthum erianthoides	M		I
Enicosanthum grandiflorum	M	D	
Enicosanthum sp.		D	
Fissistigma sp.	M	D	
Goniothalamus clemensii	M		
Goniothalamus dolichocarpus			I
Goniothalamus fasciculatus	M		
Goniothalamus gigantifolius		D	
Goniothalamus ridleyi	M		
Goniothalamus roseus	M		
Goniothalamus woodii	M	D	
Marsypopetalum pallidum		D	
Meiogyne virgata		D	
Mezzettia havilandii	M		
Mezzettia parvifolia	M		
Miliusa micropoda		D	

Species	M	D	I
Mitrephora humilis	M		
Mitrephora korthalsiana			I
Mitrephora sp.		D	
Monocarpia marginalis		D	
Neouvaria acuminatissima	M	D	
Neouvaria foetida			I
Orophea corymbosa		D	
Orophea kostermansiana		D	
Orophea myriantha	M	D	
Orophea sarawakensis		D	
Phaeanthus crassipetalus		D	
Phaeanthus laxiflora	M		
Phaenthus sp.		D	
Polyalthia bullata	M		
Polyalthia canangioides	M		
Polyalthia cauliflora	M	D	I
Polyalthia cauliflora var. *beccarii*		D	
Polyalthia chrysotricha			I
Polyalthia congesta	M	D	
Polyalthia glauca	M		
Polyalthia insignis	M	D	I
Polyalthia lateriflora	M		
Polyalthia longipes			I
Polyalthia microtus	M	D	
Polyalthia rumphii		D	
Polyalthia sumatrana	M	D	
Polyalthia xanthopetala		D	
Polyalthia aff. *xanthopetala*			I
Popowia odoardoi	M	D	
Popowia pisocarpa	M	D	
Popowia sp.			I
Pseuduvaria reticulata		D	
Pseuduvaria sp.	M		
Sageraea lanceolata	M		
Uvaria grandiflora		D	
Uvaria littoralis		D	
Uvaria ovalifolia	M		
Uvaria rufa		D	
Uvaria sorsogonensis	M	D	
Xylopia dehiscens	M		
Xylopia elliptica	M		
Xylopia ferruginea	M		
Xylopia stenopetala	M		

APOCYNACEAE

Species	M	D	I
Alstonia angustifolia	M		
Alstonia iwahigensis		D	
Alyxia pilosa	M		
Alyxia sp.		D	
Anodendron gradilis	M		
Chilocarpus beccarianus	M		
Chilocarpus vernicosus		D	
Ichnocarpus frutescens		D	
Ichnocarpus serpyllifolius		D	
Kopsia dasyrachis		D	
Kopsia sp.	M		
Leuconotis eugenifolia		D	
Microchites serpyllifolius			I
Parameria polyneura		D	
Parsonsia philippinensis		D	
Rauvolfia sumatrana		D	
Tabernaemontana macrocarpa		D	
Tabernaemontana pandacaqui		D	
Tabernaemontana pauciflora	M	D	I
Urceola brachysepala			I
Urceola laevis	M		

Species	M	D	I
Urnularia lanceolata	M		
Willughbeia coriacea	M		
Willughbeia firma		D	

AQUIFOLIACEAE

Species	M	D	I
Ilex clemensiae	M		
Ilex clethriflora		D	
Ilex cymosa	M		
Ilex glomerata	M		
Ilex oppositifolia		D	
Ilex revoluta	M		
Ilex triflora	M		
Ilex wallichii	M		

ARACEAE

Species	M	D	I
Aglaonema schottianum		D	I
Alocasia cuprea	M		
Alocasia lowii		D	
Alocasia macrorrhiza		D	
Amorphophallus sp.	M		
Anadendrum microstachyum		D	
Anadendrum sp.	M	D	
Homalomena sp.	M		I
Piptospatha sp.		D	
Pothos beccarianus			I
Pothos scandens		D	
Pothos sp.	M		
Rhaphidophora korthalsii		D	
Rhaphidophora sp.	M		
Schismatoglottis lingua	M		
Schismatoglottis motleyana	M		
Schismatoglottis mutata		D	
Schismatoglottis silamensis		D	
Schismatoglottis trifasciata	M		
Schismatoglottis wallichii		D	
Scindapsus borneensis	M		
Scindapsus pictus	M		
Scindapsus rupestris	M		

ARALIDIUACEAE

Species	M	D	I
Aralidium pinnatifidium		D	

ARALIDIACEAE

Species	M	D	I
Arthrophyllum sp.	M		
Osmoxylon borneense	M		
Schefflera elliptica	M		
Schefflera petiolosa	M		
Schefflera ridleyi	M		
Schefflera trineura	M		

ARAUCARIACEAE

Species	M	D	I
Agathis borneensis	M		
Agathis kinabaluensis	M		
Agathis lenticula	M		

ARISTOLOCHIACEAE

Species	M	D	I
Aristolachia sp.	M		
Thottea cf. *triserialis*	M		

ASCLEPIADACEAE

Species	M	D	I
Asclepias currassavica		D	
Dischidia bengalensis	M		
Dischidia hirsuta	M		
Dischidia nummularia		D	
Hoya campanulata	M		
Hoya coronaria	M		
Hoya latifolia	M		
Hoya multiflora		D	
Tylophora tenuis	M		

BEGONIACEAE

	M	D	I
Begonia berhamania	M		
Begonia beryllae	M		
Begonia aff. *densinesis* (*Gregson et al.* 122)		D	
Begonia cf. *limii*	M		
Begonia keeana	M		
Begonia postarii		D	
Begonia queritziana	M	D	
Begonia species A (*Gregson et al.* 125)		D	
Begonia species B (*Gregson et al.* 124)		D	
Begonia species C (*Gregson et al.* 85)		D	
Begonia species D (*Gregson et al.* 92)		D	
Begonia species F (*Gregson et al.* 126)		D	

BIGNONIACEAE

	M	D	I
Oroxylon indicum	M		
Radermachera pinnata subsp. *accuminata*		D	

BOMBACACEAE

	M	D	I
Durio acutifolius	M		I
Durio grandiflorus	M		I
Durio graveolens	M	D	
Durio griffithii	M	D	
Durio cf. *kinabaluensis*	M		
Durio kutejensis	M		
Durio lanceolatus	M		I
Durio oxleyanus	M		
Neesia strigosa	M		
Neesia synandra	M	D	

BORAGINACEAE

	M	D	I
Heliotropium indicum	M		
Pteleocarpa lamponga	M		

BUDDLEJACEAE

	M	D	I
Buddleja asiatica		D	

BURMANNIACEAE

	M	D	I
Burmannia longifolia	M		

BURSERACEAE

	M	D	I
Canarium asperum		D	
Canarium decumanum	M		
Canarium denticulatum subsp. *denticulatum*	M	D	
Canarium kinabaluensis	M		
Canarium littorale	M		
Canarium odontophyllum	M	D	
Canarium patentinervium	M		
Dacryodes incurvata	M		
Dacryodes laxa	M		
Dacryodes longifolia	M		
Dacryodes rostrata	M	D	
Dacryodes rubiginosa	M		
Dacryodes rugosa	M		
Dacryodes rugosa var. *rugosa*	M		
Dacryodes rugusa var. *virgata*	M	D	
Santiria grandiflora	M		
Santiria laevigata	M		
Santiria cf. *oblongifolia*	M		

CAPPARACEAE

	M	D	I
Capparis sp.		D	
Cleome rutidosperma		D	
Crateva magna		D	
Crateva religosa		D	

CAPRIFOLIACEAE

	M	D	I
Viburnum sambucinum	M		

CASAURINACEAE

	M	D	I
Gymnostoma nobile	M		
Gymnostoma sumatrana	M	D	

CELASTRACEAE

	M	D	I
Bhesa paniculata	M		
Cassine kochinchinensis	M		
Cassine viburnifolia	M		
Euonymus castaneifolius	M		
Euonymus javanicus		D	
Lophopetalum beccarianum	M	D	I
Lophopetalum glabrum			I
Lophopetalum javanicum		D	I
Lophopetalum multinervium	M		
Lophopetalum rigidum		D	
Lophopetalum subovatum	M		I
Microtropis kinabaluensis	M		
Microtropis platyphylla	M		
Microtropis sabahensis	M		
Salacia leucoclada		D	

CHLORANTHACEAE

	M	D	I
Chloranthus erectus		D	

CHRYSOBALANCEAE

	M	D	I
Atuna cordata		D	
Atuna racemosa subsp. *excelsa*	M	D	
Atuna racemosa subsp. *racemosa*		D	
Licania splendens		D	
Maranthes corymbosa		D	
Parastemon urophyllus	M		
Parinari gigantea	M		
Parinari oblongifolia	M		
Parinari parva		D	
Parinari rigida		D	

CLETHRACEAE

	M	D	I
Clethra canescens var. *clementis*	M		
Clethra pachyphylla	M		

CLUSIACEAE

	M	D	I
Calophyllum blancoi	M		
Calophyllum bursicolum	M		
Calophyllum coeletryi	M		
Calophyllum cordata	M		
Calophyllum depressinervosum	M		
Calophyllum gracilipes	M	D	
Calophyllum griseum	M		
Calophyllum nodosum	M	D	
Calophyllum obliquinervium		D	
Calophyllum pyriforme		D	
Calophyllum soulattri	M		
Calophyllum teysmannii	M		
Calophyllum venulosum		D	
Calophyllum wallichanum var. *incrassatum*	M		
Garcinia bancana		D	
Garcinia benthamiana	M		D

	M	D	I
Garcinia cf. *celebica*	M		
Garcinia desrousseauxii	M		
Garcinia diospyrifolia	M		
Garcinia forbesii	M	D	I
Garcinia gaudichandii	M		I
Garcinia gracilipes	M		
Garcinia mangostana	M		
Garcinia miquelii	M		
Garcinia multinerva	M		
Garcinia nodosum	M		
Garcinia parvifolia	M	D	
Garcinia penangiana			I
Garcinia ramiflora	M		
Garcinia venulosa	M		
Cratoxylum arborescens	M	D	I
Cratoxylum cochincinensis	M	D	I
Cratoxylum formosum	M	D	I
Cratoxylum sumatranum	M	D	I
Kayea imbricata		D	
Mesua borneensis	M		
Mesua calciola			I
Mesua hexapetala			I
Mesua macrantha	M		
Mesua planigemma		D	

COMBRETACEAE

	M	D	I
Combretum sundaicum			I
Terminalia citrina		D	
Terminalia feotidissima	M		

COMMELINACEAE

	M	D	I
Amischotolype glabrata		D	
Amischotolype griffithii	M	D	
Amischotolype marginata		D	
Amischotolype mollissima			I
Belosynapsis ciliata		D	
Commelina mudiflora	M		
Forrestia sp.	M		
Pollia subumbellata		D	
Pollia thyrsiflora		D	
Pollia sp.	M	D	I
Tricarpelema philippense		D	I
Tricarpelema pumilum		D	

COMPOSITAE

	M	D	I
Adenostemma macrophylla	M		
Adenostemma sp.		D	
Ageratum balsamifera		D	
Ageratum conyzoides		D	
Blumea balsamifera	M	D	
Chromolaena odorata		D	
Crassocephalum crepidioides		D	
Emilia sp.	M		
Erechtites sp.		D	
Eupatorium odoratum		D	
Gynura procumbens	M		
Senecio sp.	M		
Sphagneticola trilobata		D	
Vernonia arborea		D	

CONNARACEAE

	M	D	I
Agelaea borneensis	M	D	
Agelaea insignis			I
Agelaea tinervis	M	D	I
Connarus euphlebius	M		
Connarus grandis		D	
Connarus odoratus		D	

CONVOLVULACEAE

	M	D	I
Erycibe borneensis	M	D	I
Erycibe grandifolia		D	
Erycibe impressa			I
Erycibe stenophylla			I
Gymnostemma pentaphylla			I
Gymnostemma sp.		D	
Ipomoea cairica		D	
Jacquemontia tomentella var. *micrantha*		D	
Merremia borneensis		D	
Merremia clemensiana		D	
Merremia gracile		D	
Merremia korthalsiana		D	
Merremia peltata		D	

CORNACEAE

	M	D	I
Mastixia rostrata subsp. *caudatifolia*	M	D	I

COSTACEAE

	M	D	I
Costus globosus		D	
Costus paradoxus			I
Costus speciosus	M		

CRYPTERONIACEAE

	M	D	I
Axinandra coriacea	M	D	
Crypteronia grifthii	M		

CUCURBITACEAE

	M	D	I
Alsomitra sp.	M		
Beccariana sp.	M		
Benincasa sp.	M		
Gymnopetalum chinense	M		I
Gymnostemma sp.	M		I
Hodgsonia macrocarpa	M		I
Mimordica cochinchinensis	M	D	
Momordica denticulata			I
Siraitia sp.	M		
Trichosanthes beccariana	M		
Trichosanthes elmeri			I
Trichosanthes intermedia	M		
Trichosanthes montana			I
Trichosanthes pendula	M	D	
Trichosanthes postarii	M		I
Trichosanthes pubera	M		
Trichosanthes quinquanggulata	M		
Trichosanthes sepilokensis	M		
Zehneria marginata	M		

CUNONIACEAE

	M	D	I
Weinmannia fraxinea	M		

CYPERACEAE

	M	D	I
Carex saturata	M		
Carex sp.		D	
Cyperus compactus		D	
Cyperus diffusus	M		
Cyperus iria		D	
Cyperus kyllingia		D	
Cyperus laxus subsp. *macrostachyus*			I
Cyperus luzulae	M	D	
Cyperus polystachyos		D	
Cyperus procerus		D	
Eleocharis retroflexa	M		
Eleocharis sp.			I
Fimbristylis dichotoma		D	

Species	M	D	I
Fimbristylis dura	M		
Fimbristylis miliacea		D	
Kyllinga sp.	M		
Mapania cuspidata	M	D	
Mapania cuspidata var. petiolta		D	
Mapania graminea		D	
Mapania meditensis		D	
Mapania palustris var. palutris	M	D	
Mapania richardsii	M		
Mapania urceolata	M		
Mapania sp.			I
Mariscus cyperinus			I
Mariscus microcephalus		D	
Oreobolus radicans	M		
Paramapania radiatus			I
Rhynchospora corymbosa	M		
Scleria ciliaris	M		
Scleria motleyi	M		
Scleria purpurascens	M	D	I
Trichophorum sp.	M		

DAPHNIPHYLLACEAE

Species	M	D	I
Daphniphyllum laurinum	M		

DATISCACEAE

Species	M	D	I
Octomeles sumatrana		D	

DILLENIACEAE

Species	M	D	I
Dillenia beccariana	M		
Dillenia borneensis	M		
Dillenia excelsa	M	D	
Dillenia excelsa var. pubescens		D	
Dillenia suffruticosa			I
Dillenia sumatrana		D	
Tetracera akara	M	D	
Tetracera korthalsii	M		
Tetracera scandens	M	D	

DIOSCOREACEAE

Species	M	D	I
Dioscorea bulbifera			I
Dioscorea sumatrana		D	
Dioscorea sp.	M	D	

DIPTEROCARPACEAE

Species	M	D	I
Anisoptera costata		D	I
Anisoptera sp.	M		
Cotylelobium melanoxylon			I
Dipterocarpus acutangulus	M	D	I
Dipterocarpus applanatus	M		I
Dipterocarpus candiferus	M	D	I
Dipterocarpus confertus	M		I
Dipterocarpus crinitus	M		
Dipterocarpus elongatus			I
Dipterocarpus eurynchus	M		
Dipterocarpus gracilis	M		I
Dipterocarpus kerii		D	I
Dipterocarpus kunstleri			I
Dipterocarpus lowii	M		
Dipterocarpus megacarpus			I
Dipterocarpus stellata subsp. parvus	M		
Dipterocarpus tempehes			I
Drybalanops beccarii			I
Dryobalanops keithii			I
Drybalanops lanceolata	M	D	I
Hopea aequalis	M		
Hopea beccariana	M		
Hopea bracteata	M		I
Hopea dryobalanoides		D	
Hopea ferruginea	M		I
Hopea nervosa	M	D	I
Hopea pentanervia			I
Hopea sangal	M		I
Hopea wyatt-smythii		D	
Parashorea malaanonan	M	D	I
Parashorea tomentella	M	D	I
Shorea acuminatissima	M		
Shorea agamii subsp. agamii	M		I
Shorea almom	M		I
Shorea amplexicaulis			I
Shorea andulensis	M		I
Shorea angustifolia	M	D	
Shorea argentifolia	M		I
Shorea asahii			I
Shorea atrinervosa	M	D	I
Shorea beccariana			I
Shorea bracteolata	M		
Shorea confusa		D	I
Shorea coriacea	M		
Shorea faguetiana	M	D	I
Shorea falciferoides subsp. glaucescens	M		
Shorea fallax	M	D	I
Shorea ferruginea	M		
Shorea flaviflora			I
Shorea cf. flemmichii	M		
Shorea foxworthyi	M		I
Shorea gibbosa	M	D	I
Shorea glaucescens			I
Shorea gratissima	M		
Shorea guiso		D	
Shorea hopeifolia	M		
Shorea hypoleuca		D	
Shorea johorensis	M	D	I
Shorea kunstleri			I
Shorea laevis	M		I
Shorea leprosula	M	D	I
Shorea macrophylla	M		I
Shorea macrophylla x subsp. pinanga			I
Shorea macroptera			I
Shorea maxwelliana			I
Shorea mecistopteryx	M		I
Shorea monticola			I
Shorea multiflora	M		I
Shorea obscura	M		
Shorea ochracea			I
Shorea oleosa	M		
Shorea cf. oleuca	M		
Shorea ovalis	M		
Shorea ovalis subsp. ovalis			I
Shorea parvistipulata	M	D	I
Shorea parvifolia subsp. parvifolia		D	I
Shorea patoiensis	M		I
Shorea pauciflora	M	D	I
Shorea pilosa	M	D	
Shorea pinanga	M		
Shorea platycarpa	M		
Shorea platyclados	M		
Shorea rubra			I
Shorea scrobiculata	M		
Shorea seminis			I
Shorea smithiana	M	D	I
Shorea superba	M	D	
Shorea symingtonii		D	I
Shorea venulosa	M		I

	M	D	I
Shorea waltoni	M		I
Shorea xanthophylla			I
Vatica albiramis		D	I
Vatica dulitensis	M	D	I
Vatica oblongifolia	M	D	I
Vatica rassak	M		
Vatica sarawakensis		D	
Vatica umbonata		D	

DRACAENACEAE

	M	D	I
Dracaena angustifolia	M	D	I
Dracaena elliptica	M	D	

EBENACEAE

	M	D	I
Diospyros andamanica		D	
Diospyros buxifolia	M		
Diospyros cauliflora	M	D	
Diospyros curranii	M		I
Diospyros curraniopsis	M		
Diospyros daemona		D	
Diospyros densa	M		
Diospyros diepenhorstii		D	
Diospyros elliptifolia	M	D	I
Diospyros euphlebia		D	
Diospyros foxworthyi	M		
Diospyros frutescens		D	
Diospyros fusiformis	M	D	
Diospyros korineii	M		
Diospyros laevigata	M		
Diospyros lanceifolia	M		
Diospyros macrophylla	M	D	
Diospyros mindanaensis			I
Diospyros nitida	M	D	
Diospyros oligantha			I
Diospyros penibukanensis	M		
Diospyros squamaefolia		D	
Diospyros subrhomboidea			I
Diospyros sumatrana	M	D	
Diospyros toposia var. *toposioides*		D	I
Diospyros tuberculata		D	

ELAEOCARPACEAE

	M	D	I
Elaeocarpus beccarii		D	
Elaeocarpus brunnescens	M		
Elaeocarpus clementis var. *clementis*			I
Elaeocarpus ferrugineus	M		I
Elaeocarpus hullettii	M		
Elaeocarpus jugahanus	M		
Elaeocarpus knuthii subsp. *knuthii*		D	
Elaeocarpus mastersii	M		
Elaeocarpus murudensis	M		
Elaeocarpus obtusifolius	M		
Elaeocarpus pachyophrys	M		
Elaeocarpus aff. *palembanicus*	M		
Elaeocarpus petiolatus	M		
Elaeocarpus stipularis		D	
Elaeocarpus stipularis var. *brevipes*			I

ERICACEAE

	M	D	I
Costera endertii		D	
Dendrotrophe varians		D	
Diplycosia barbigera	M		
Diplycosia chrysothrix	M		
Diplycosia heterophylla	M		
Diplycosia heterophylla var. *latifolia*	M		
Diplycosia sp.	M		
Diplycosia cf. *microphylla*	M		

	M	D	I
Gaultheria sp.	M		
Rhododendron bagobonum	M		
Rhododendron borneense		D	
Rhododendron borneense subsp. *villosum*	M		
Rhododendron burttii	M		
Rhododendron crassifolium	M		
Rhododendron cuneifolium	M		
Rhododendron durionifolium	M		
Rhododendron durionifolium subsp. *sabahense*	M		I
Rhododendron fallacinum	M		
Rhododendron javanicum		D	
Rhododendron javanicum subsp. *brookeanum*	M		
Rhododendron javanicum subsp. *cladotrichum*		D	
Rhododendron javanicum subsp. *cockburnii*	M		
Rhododendron javanicum subsp. *gracile*	M		
Rhododendron longiflorum subsp. *longiflorum*	M		
Rhododendron longiflorum subsp. *subcordatum*	M		I
Rhododendron malayanum	M	D	
Rhododendron micromalayanum	M		
Rhododendron nervulosum	M		
Rhododendron orbiculatum	D		
Rhododendron praetervisum	M		
Rhododendron stapfianum	M		
Rhododendron suaveolens	M		
Vaccinium bancanum	M		
Vaccinium ceridifolium	M	D	
Vaccinium claoxylon	M		
Vaccinium clementis	M		
Vaccinium coriaeum	M		
Vaccinium pachydermum	M		
Vaccinium phillyreoides	M		
Vaccinium sp.			I

ERYTHROXYLACEAE

	M	D	I
Erythroxylum cuneatum forma *sumatranum*	M	D	

EUPHORBIACEAE

	M	D	I
Agrostistachys borneensis	M		
Agrostistachys leptostachys	M		
Agrostistachys longifolia	M		
Antidesma leucopodum	M		
Antidesma montis-silam		D	I
Antidesma neurocarpum var. *hosei*		D	
Antidesma neurocarpum var. *neurocarpum*	M	D	I
Antidesma puncticulatum		D	
Antidesma riparium var. *riparium*		D	
Antidesma stipulare	M	D	
Antidesma tomentosum var. *tomentosum*	M	D	I
Antidesma venenosum	M	D	
Aporosa acuminatissima	M		
Aporosa aurea	M		
Aporosa confusa	M	D	
Aporosa elmeri	M	D	I
Aporosa falcifera	M	D	
Aporosa frutescens	M	D	I

Species	M	D	I
Aporosa grandistipulata	M	D	
Aporosa lucida	M		I
Aporosa lunata			I
Aporosa nigricans		D	
Aporosa nitida	M	D	
Aporosa prainiana		D	
Aporosa subcaudata	M	D	
Baccaurea javanica	M		
Baccaurea kunstleri			I
Baccaurea lanceolata	M	D	
Baccaurea marcocarpa		D	I
Baccaurea marcophylla	M		
Baccaurea minor	M		
Baccaurea racemosa		D	I
Baccaurea stipulata	M	D	
Baccaurea sumatrana	M	D	I
Baccaurea tetrandra	M	D	I
Baccaurea trigonocarpa	M		I
Blumeodendron kurzii	M		I
Blumeodendron tokbrai	M	D	
Borneodendron aenigmaticum	M	D	
Breynia conorata	M	D	
Bridelia griffitii var. cinnamomea		D	
Cephalomappa beccariana	M		
Cephalomappa lepidotula	M		
Chaetocarpus castanocarpus	M		
Claoxylon sp.		D	
Cleistanthus baramicus	M		
Cleistanthus everetti		D	
Cleistanthus glaber		D	
Cleistanthus gracilis	M		
Cleistanthus hirsutulus		D	
Cleistanthus maingayi			I
Cleistanthus megacarpus	M		
Cleistanthus myrianthus	M		
Cleistanthus cf. oblongatus	M		
Cleistanthus pedicellatus		D	
Cleistanthus podopyxis		D	
Cleistanthus sumatranus	M		I
Cleistanthus venosus			I
Croton argyratus	M		
Croton ensifolius	M		
Croton griffithii		D	
Croton oblongifolius	M		
Croton rheophyticus	M		
Dimorphocalyx muricatus	M	D	
Drypetes eriocarpa	M	D	
Drypetes gracilipes	M		
Drypetes kikir	M		
Drypetes longifolia	M	D	I
Drypetes macrostigma	M		
Drypetes subcubica	M	D	
Endospermum sp.		D	
Endospermum diadenum	M		
Endospermum malaccensis	M		
Endospermum marcophylla	M		
Endospermum peltatum	M		
Fahrenheitia pendula	M	D	I
Galearia fulva	M		
Glochidion borneense			I
Glochidion calospermum	M		
Glochidion elmeri	M	D	
Glochidion hypoleucum	M	D	
Glochidion lanceilimbum		D	
Glochidion lanceisepalum		D	
Glochidion lanceolatum		D	
Glochidion lutescens	M	D	
Glochidion obscurum	M	D	I
Glochidion pubicapsa		D	
Glochidion rubrum	M	D	I
Glochidion wallichianum	M		
Homalanthus populneus	M		
Homonoia riparia	M		
Koilodepas laevigatum	M	D	I
Koilodepas longifolium		D	I
Koilodepas pectanatum	M	D	
Macaranga beccariana	M		
Macaranga brevipetiolata			I
Macaranga depressa		D	
Macaranga gigantifolis	M		
Macaranga hypoleuca	M	D	I
Macaranga indistincta		D	
Macaranga kinabaluensis	M	D	
Macaranga lakeyi	M		
Macaranga lowii	M		
Macaranga lowii var. kostermansii			I
Macaranga macrostachys	M		
Macaranga motleyana		D	
Macaranga pearsonii	M		
Macaranga penangensis	M		
Macaranga prunosa	M		
Macaranga puberula	M		
Macaranga recurvata	M		
Macaranga repando-denata	M		
Macaranga tribola	M		
Macaranga winkleri	M	D	
Macaranga wrayi	M		
Mallotus caudatus	M		
Mallotus eucaustus		D	I
Mallotus griffithianus	M		
Mallotus korthalsii	M		
Mallotus lackeyi		D	I
Mallotus marcostachyus		D	
Mallotus miquelianus		D	I
Mallotus muticus	M		
Mallotus oblongifolius	M		
Mallotus paniculatus		D	
Mallotus peltatus		D	
Mallotus penangensis	M	D	I
Mallotus stercularis	M		
Mallotus stipularis	M	D	
Mallotus wrayi	M	D	
Melanolepis multiglandulosa	M		
Moultonianthus leembruggianus	M		
Neoscortechinia forbesii	M	D	
Neoscortechinia philippinensis		D	I
Omphalea bracteata	M		I
Phyllanthus balgooyi	M		
Phyllanthus lamprophyllus	M		
Phyllanthus urinaria			I
Pimeleodendron griffithianum	M		
Ptychopyxis arborea	M		
Ptychopyxis bacciformis	M		
Ryporasa sp.		D	
Sauropus rhamnoides	M		
Sauropus sp.		D	
Spathiostemon javanensis	M	D	
Suregada glomerulata	M	D	
Syndyophyllum excelsum subsp. occidentale		D	
Trigonostemon elmeri		D	I
Trigonostemon ionthocarpus			I
Trigonostemon malayana	M		

FAGACEAE

	M	D	I
Castanopsis hypophoenicea	M	D	I
Castanopsis motleyana	M		
Castanopsis psilophylla	M		
Lithocarpus bancanus		D	
Lithocarpus bennettii	M		
Lithocarpus cantleyanus	M		
Lithocarpus caudatifolius	M		
Lithocarpus clementianus	M		
Lithocarpus confertus	M		
Lithocarpus conocarpus	M		
Lithocarpus dasystachyus	M		
Lithocarpus elegans	M	D	
Lithocarpus ewyckii	M		
Lithocarpus gracilis	M	D	
Lithocarpus hallieri	M		
Lithocarpus hatusimae	M		
Lithocarpus havilandii	M		
Lithocarpus leptogyne	M	D	I
Lithocarpus lucidus	M		
Lithocarpus meijeri	M		
Lithocarpus nieuwenhuisii	M	D	
Lithocarpus revolutus	M		
Lithocarpus tawaiensis	M		
Quercus argentata	M	D	
Quercus lowii	M		
Quercus sumatrana	M		
Quercus treubiana	M		
Quercus valdinervosa	M		
Trigonobalanus verticillatus	M		

FLACOURTIACEAE

	M	D	I
Casearia capitellata	M		
Casearia rugulosa	M	D	I
Casearia tuberculata		D	
Flacourtia rukam	M		
Homalium panayanum		D	
Hydnocarpus anomala	M		
Hydnocarpus borneensis	M	D	I
Hydnocarpus calophylla	M		
Hydnocarpus polypetala	M	D	
Hydnocarpus sumatrana	M		
Hydnocarpus woodii	M		
Pangium edule	M		
Ryparosa acuminata	M	D	I
Ryparosa baccaureoides	M		
Ryparosa hullettii	M	D	I
Ryparosa kostermanii		D	

FLAGELLARIACEAE

	M	D	I
Flagellaria indica		D	

GENTIANACEAE

	M	D	I
Cotylanthera sp.		D	

GESNERIACEAE

	M	D	I
Aeschynanthus albidus	M		
Aeschynanthus curtisii			I
Aeschynanthus maquiticus	M		
Aeschynanthus tricolor	M	D	
Agalmyla sp.	M	D	
Cyrtandra angularis		D	I
Cyrtandra angularius	M		
Cyrtandra areolata	M		
Cyrtandra chrysea	M		
Cyrtandra gibbsiae		D	
Cyrtandra cf. kermesina	M		
Cyrtandra longicarpa	M		
Cyrtandra cf. multibracteata	M		
Cyrtandra oblongifolia		D	
Cyrtandra sarawakensis		D	
Cyrtandra simplex		D	
Cyrtandra warburgiana		D	
Didymocarpus crocea	M		
Didymocarpus cf. hispida	M		
Didymocarpus sp.			I
Epithema carnosum		D	
Epithema dolichopodum	D		
Gesneria sp.	M		
Henckelia amoenus	M	D	
Henckelia gracilipes	M		
Henckelia lanceolata			I
Hexatheca dolichopoda		D	
Monophyllaea merrilliana		D	
Rhynchoglossum klugioides		D	
Rhynchoglossum medusothrix		D	

GNETACEAE

	M	D	I
Gnetum diminutum	M		
Gnetum gnemon var. brunonianum	M		
Gnetum gnemonoides		D	
Gnetum leptostachyum		D	
Gnetum neglectum	M		
Gnetum sp.			I

GOODENIACEAE

	M	D	I
Scaevola micrantha		D	

GRAMINEAE

	M	D	I
Bambusa sp.	M		
Centotheca lappacea	M	D	
Cynodon dactylon	M		
Cytococcum accrescens	M	D	
Cytococcum oxyphylkim	M		
Cytococcum patens	M	D	
Dinochloa colona		D	
Dinochloa darvelana	M	D	
Dinochloa scandens	M		
Dinochloa sublaevigata		D	
Echinochloa procera	M		
Eleusine indica		D	
Eragrostis unioloides		D	
Garnotia acutighuma	M		
Ichnenthus cylindrica	M		
Ichnenthus pallens var. pallens	M		
Imperata conferta		D	
Kinabaluchloa sp.	M		
Leptochloa panicea	M		
Lophatherum gracile	M		
Oplismenus compositus	M	D	
Oplismenus hirtellus	M		
Oryza sp.	M		
Paspalum conjugatum	M	D	
Paspalum longifolium	M		
Paspalum scrobiculatum	M		
Paspalum virgatum	M		
Pogonatherum crinitum	M	D	I
Racemobambos hirsuta		D	
Schizostachyum cf. longispiculatum	M		
Scolochloa ureolata	M	D	
Thysanolaena maxima		D	
Yushania tesselata	M		

HANGUANACEAE

	M	D	I
Hanguana major		D	
Hanguana malayana	M	D	

HYPOXIDACEAE

	M	D	I
Curculigo latifolia	M	D	
Curculigo racemosa			I

ICACINACEAE

	M	D	I
Gonocaryum cognatum		D	
Gonocaryum macrophyllum		D	
Iodes philippinensis	M		
Sarcostigma paniculata	M	D	
Sleumeria auriculata		D	I
Stemonurus grandifolius	M		

ILLICIACEAE

	M	D	I
Illicium kinabaluensis	M		
Illicium stapfii	M		

JOINVILLEACEAE

	M	D	I
Joinvillea sp.	M	D	

JUGLANDACEAE

	M	D	I
Engelhardia danumensis		D	
Engelhardia serrata	M		

JUNCACEAE

	M	D	I
Juncus sp.	M		

LAMIACEAE

	M	D	I
Callicarpa candicans	M	D	
Callicarpa involucrata		D	
Callicarpa longifolia	M	D	I
Callicarpa pentandra		D	
Clerodendrum adenophysum		D	
Clerodendrum dispariifolium		D	
Clerodendrum pygmaeumM		D	
Congea tomantosa		D	
Gomphostemma curtisii		D	
Gomphostemma hirsuta		D	
Gomphostemma microcalyx	M		
Hyptis capitata		D	
Leucas aspera		D	
Plectranthus sp.		D	
Petraeovitex ternata	M		
Petraeovitex sp.		D	
Premna oblongata		D	
Premna serratifolia	M	D	
Premna thrichostoma		D	I
Sphenodesma trifolia		D	
Teijsmanniodendron bogoriense		D	I
Teijsmanniodendron glabrum	M		
Teijsmanniodendron holophyllum	M	D	
Teijsmanniodendron hollrungii		D	
Teijsmanniodendron pteropodum		D	
Teijsmanniodendron sarawakanum		D	
Teijsmanniodendron simplicifolium	M	D	I
Teijsmanniodendron simplicioides		D	
Teijsmanniodendron sinclairii		D	
Teijsmanniodendron subspicatum		D	
Teijsmanniodendron unifoliolatum		D	
Vitex pinnata		D	
Vitex vestita		D	

LAURACEAE

	M	D	I
Actinodaphne borneensis	M		
Actinodaphne diversifolia	M		
Actinodaphne oleifolia	M		
Actinodaphne pruinosa	M		I
Actinodaphne venosa	M		
Alseodaphne insignis	M		
Alseodaphne oblanceolata	M		
Alseodaphne rubiginosa	M		
Beilschmiedia assamica	M		
Beilschmiedia glabra	M		
Beilschmiedia jacobsii	M		
Beilschmiedia micrantha	M		
Beilschmiedia tawaensis	M		
Caryodaphnopsis tonkenensis	M	D	
Cinnamomum griffithii	M		
Cinnamomum javanicum	M		
Cinnamomum racemosum	M		
Cryptocarya cagayanensis	M		
Cryptocarya densiflora	M		
Cryptocarya pulchrianervia	M		
Dehaasia caesia	M		
Dehaasia gigantocarpa		D	
Dehaasia incrassata	M		
Endiandra macrophylla	M	D	I
Eusideroxylon zwageri	M	D	I
Lindera bibracteata	M		
Lindera caesa var. rufa	M		
Litsea accedens var. accedens		D	I
Litsea accedens	M		
Litsea andreana var. diwolii		D	
Litsea brachystachya	M		
Litsea calicarpa	M		
Litsea caulocarpa		D	
Litsea crassifolia	M		
Litsea elliptica	M		
Litsea ellipticacea	M		
Litsea fulva	M	D	I
Litsea globularia	M		
Litsea lancifolia var. grandifolia	M	D	
Litsea lancifolia var. iliaspaiei		D	
Litsea machilifolia		D	
Litsea odorifera	M		
Litsea ochracea		D	
Litsea oppositifolia	M		
Litsea resinosa	M		
Litsea sessiliflora	M	D	
Litsea sessiliflora var. othmanii		D	
Litsea sessilis	M	D	
Litsea subumbelliflora		D	
Neolitsea auricolor		D	
Neolitsea cassia	M		
Neolitsea zeylanica	M		I
Notaphoebe sarawakensis	M		
Persea bancana	M		
Phoebe macrophylla	M		

LECYTHIDACEAE

	M	D	I
Barringtonia curranii		D	
Barringtonia gigantostachya var. megistophylla		D	
Barringtonia lanceolata	M	D	
Barringtonia macrostachya		D	
Barringtonia sarcostachys	M		
Planchonia brevistipitata		D	I

LEEACEAE

	M	D	I
Leea aculeata	M	D	
Leea indica	M	D	

LEGUMINOSAE

	M	D	I
Adenanthera pavonina	M		
Afzelia rhomboidea			I
Albizia singularis	M		
Albizia splendens	M		
Archidendron borneense	M		

	M	D	I
Archidendron clypearia var. *casai*	M		
Archidendron cockburnii		D	
Archidendron ellipticum			I
Archidendron fagifolium		D	
Archidendron microcarpum			I
Archidendron rostulatum			I
Caesalpinia latisiliqua	M	D	
Caesalpinia oppositifolia		D	
Caesalpinia parviflora		D	
Caesalpinia sappan	M		
Calleria nieuwenhuisii			I
Calleria sp.	M		
Cassia tora		D	
Crudia ornata	M	D	
Crudia reticulata	M	D	
Cynometra ramiflora var. *ramiflora*		D	
Dalbergia pseudo-sossoo		D	
Derris sp.	M		
Dialium indum var. *indum*	M	D	
Dialium kunstleri	M		
Dialium platysepalum	M		I
Entada sp.	M		
Fordia brachybotrys		D	
Fordia coriacea		D	
Fordia filipes		D	I
Fordia gibbsiae		D	
Fordia cf. *johorensis*			I
Fordia splendidissima	M	D	
Koompassia excelsa	M	D	I
Koompassia malaccensis	M	D	I
Mastersia bakeri		D	
Mastersia borneensis		D	
Millettia cf. *vasta*	M		
Mucuna biplicata	M		
Mucuna cf. *hainanensis* subsp. *multilamellata*		D	
Mucuna mikilii		D	
Parkia javanica	M		
Parkia jiringa	M		
Parkia speciosa	M		
Phanera (*Bauhinia*) *diptera*	M		
Phanera (*Bauhinia*) *integrifolia* subsp. *cumingiana*		D	
Phanera (*Bauhinia*) *kockiana*	M	D	
Phanera (*Bauhinia*) *semibifida* var. *semibifida*		D	
Phanera (*Bauhinia*) *sylvani*		D	
Peltophorum racemosum	M	D	
Saraca declinata	M	D	I
Sindora irpicina	M		
Sindora velutina	M		
Spatholobus gyrocarpus	M		
Spatholobus latibractea	M		
Spatholobus macropterus		D	

LILIACEAE

	M	D	I
Dianella ensifolia	M		

LINACEAE

	M	D	I
Ctenolophon parvifolius	M		
Indorouchera sp.	M		
Ixonanthes reticulata	M		

LOGANIACEAE

	M	D	I
Fagraea blumei	M		
Fagraea cuspidata	M	D	
Fagraea involucrata	M		
Fagraea kinabaluensis		D	
Fagraea kuminii	M		
Fagraea macroschypa	M	D	
Fagraea renae		D	
Fagraea spicata	M	D	I
Fagraea splendens	M	D	
Mitrasacme sp.	M		
Strychnos ignatii	M		I

LORANTHACEAE

	M	D	I
Dendrophthoe constricta	M		
Helixanthera maxwelliana	M	D	
Helixanthera setigera		D	
Lampas elmeri		D	I
Lepeostegeres bahajensis		D	I
Lepeostegeres centiflorus		D	
Lepidaria pulchella		D	
Macrosolen cochinchinensis	M	D	

LOWIACEAE

	M	D	I
Orchidantha quadricolor		D	

LYTHRACEAE

	M	D	I
Lagerstroemia speciosa		D	

MAGNOLIACEAE

	M	D	I
Magnolia candollii var. *candollii*	M	D	
Magnolia candollii var. *singapurensis*		D	
Magnolia gigantifolia	M	D	
Michelia sp.	M		
Talauma craibiana	M		

MALVACEAE

	M	D	I
Abelmoschus moschantus		D	
Abutilon inducum		D	
Hibiscus surattensis		D	
Urena lobata		D	

MARANTACEAE

	M	D	I
Donax canniformis		D	I
Maranthus sp.	M		
Phacelophrynium aurantium	M	D	I
Phacelophrynium maximum	M	D	I
Phrynium capitatum			I
Phrynium pubinerve		D	
Phrynium villosulum	M		
Stachyphrynium borneense	M		I
Stachyphrynium sumatranum	M	D	I

MELASTOMATACEAE

	M	D	I
Allomorphia sp.	M		
Anerincleistus echinatus	M		
Anerincleistus macrophylla	M		
Anerincleistus setulosus	M		
Astronia sp.	M		
Blastus cogniauxii	M		
Catanthera sp.			I
Catanthera tawaoensis	M		
Clidemia hirta		D	
Creaghiella purpurea	M		
Creaghiella setosa	M		
Diplectria divaricata		D	
Diplectria glabra	M		
Diplectria glabra var. *kinabaluensis*		D	
Diplectria micrantha	M		
Diplectria viminalis		D	
Dissochaeta annulata	M		
Dissochaeta beccariana	M		

Species	M	D	I
Dissochaeta aff. *beccariana*			I
Dissochaeta bracteata		D	
Dissochaeta gracilis		D	
Dissochaeta punctulata	M		
Dissochaeta rostrata var. *porphyrocarpa*		D	I
Dissochaeta rubignosa	M		
Dissochaeta stellata		D	
Dissochaeta stellualata		D	
Dissochaeta stipularis		D	
Dissochaeta viminalis	M	D	
Driessenia inaequifolia var. *inaequifolia*		D	
Driessenia microthrix	M		
Kibessia galeata	M	D	
Kibessia korthalsia	M		
Macrolenes stellulata var. *stellulata*		D	
Medinilla alternifolia		D	
Medinilla crassifolia	M		
Medinilla danumensis		D	
Medinilla laxiflora	M	D	
Medinilla macrophylla	M	D	
Medinilla cf. *quadrifolia*	M		
Medinilla suberosa	M		
Medinilla succulenta	M		
Medinilla tawaoensis	M		
Melastoma anomala	M		
Melastoma beccarianum	M		
Melastoma laevifolia	M		
Melastoma malabathricum	M	D	I
Melastoma neccarianum	M		
Melastoma oxpora	M		
Memecylon appendiculatum	M		
Memecylon beccarianum	M		
Memecylon borneensis	M		
Memecylon costatum	M		
Memecylon edule	M		
Memecylon laevigatum	M		
Memecylon paniculatum	M		
Memecylon sp.		D	I
Ochthocharis sp.	M		
Oxyspora auriculata	M		
Oxyspora cordata	M		
Pachycentria constrcta	M	D	
Pachycentria pulverulenta	M		
Phyllagathis sp.	M		
Pternandra azurea var. *azurea*	M	D	
Pternandra cogniauxii	M		I
Pternandra coerulescens	M	D	I
Pternandra galeata		D	
Pternandra rostrata	M	D	
Sonerila borneensis	M		
Sonerila crassiusscule	M		
Sonerila kinabaluensis	M		
Sonerila sp.		D	I
Ternandra sp.		D	

MELIACEAE

Species	M	D	I
Aglaia beccarii	M		
Aglaia crassinervia	M	D	
Aglaia elliptica subsp. *elliptica*	M	D	
Aglaia elliptica subsp. *clementis*		D	
Aglaia exstipulata subsp. *brunneostellata*		D	
Aglaia forbesii	M		
Aglaia glabrata	M		
Aglaia grandis		D	
Aglaia hiernii		D	
Aglaia korthalsii		D	I
Aglaia lawii subsp. *oligocarpa*	M	D	
Aglaia laxiflora		D	
Aglaia leptantha	M		I
Aglaia leucophylla	M		
Aglaia luzoniensis	M	D	
Aglaia macrocapa		D	
Aglaia malaccensis	M		
Aglaia meliosmoides		D	I
Aglaia multinervis		D	
Aglaia odoratissima	M	D	I
Aglaia oligophylla	M	D	
Aglaia pachyphylla		D	
Aglaia palembanica	M		
Aglaia rivularis	M	D	I
Aglaia rufinervis	M	D	
Aglaia silvestris		D	
Aglaia tenuicaulis			I
Aglaia teysmanniana		D	
Aglaia tomentosa subsp. *tomentosa*	M	D	I
Aglaia tomentosa subsp. *kabaensis*	M	D	
Aphanamixis borneensis	M		I
Aphanamixis polyachya	M	D	
Chisocheton beccarianum	M		
Chisocheton cumingianus subsp. *kinabaluensis*			I
Chisocheton divergens	M		
Chisocheton erythrocarpus		D	
Chisocheton macranthus		D	
Chisocheton medusae			I
Chisocheton patens	M	D	I
Chisocheton pentandrus	M	D	I
Chisocheton pentandrus subsp. *medius*	M	D	
Chisocheton pentandrus subsp. *pauchijugus*			I
Chisocheton sarawakanus	M	D	
Dysoxylum cf. *acutangulum*	M		
Dysoxylum arborescens	M		
Dysoxylum cauliflorum	M	D	
Dysoxylum cyrtobotryum	M	D	
Dysoxylum grande	M	D	
Dysoxylum nigulosum	M		
Dysoxylum pachyrhache	M		
Dysoxylum rigidum		D	
Dysoxylum rugulosum	M	D	
Dysoxylum sp.			I
Heyna trijuga		D	
Lansium domesticum	M		I
Reinwardthiodendron humile	M	D	
Sandoricum koetjape	M		
Walsura pinnata	M	D	I
Xylocarpus granatum		D	

MELIOSMACEAE

Species	M	D	I
Meliosma sumatrana	M	D	

MENISPERMACEAE

Species	M	D	I
Coscinium fenestratum	M	D	
Fibraurea chloroleuca	M		
Haematocarpus validus	M		
Parabanea megalocarpa		D	
Pycnarrhena tumerfacta		D	
Stephania corymbosa		D	
Stephania reticulata	M	D	

MONIMIACEAE

Kibara ontusa M D

MORACEAE

Antiaris taxicaria	M		
Artocarpus anisophyllus	M		
Artocarpus dadah	M	D	
Artocarpus elasticus	M		
Artocarpus kemando	M		
Artocarpus lanceifolius	M		
Artocarpus nitidus	M		
Artocarpus odoratissimus	M		
Artocarpus primackiana		D	
Artocarpus tamaran		D	
Ficus annulata	M	D	
Ficus apiocarpa			I
Ficus aurantiacea var. *parvifolia*	M		
Ficus binnendykii	M		
Ficus cereicarpa	M		
Ficus chartacea		D	
Ficus cucurbitina	M		
Ficus cuspida	M		
Ficus delosyce	M		
Ficus deltoidea	M	D	
Ficus deltoidea var. *intermedia*		D	
Ficus depressa	M	D	
Ficus aff. *endospermifolia*	M		
Ficus fistulosa	M		
Ficus heteropleura var. *heteropleura*	M		I
Ficus lepicarpa	M	D	
Ficus leptocalama	M		
Ficus megaleia var. *subuncinata*	M		
Ficus microcarpa		D	
Ficus midotis		D	
Ficus obscura var. *obscura*	M	D	
Ficus oleifolia var. *memecylifolia*	M		
Ficus parietalis	M	D	I
Ficus pendens		D	
Ficus pisocarpa			I
Ficus punctata		D	
Ficus retusa			I
Ficus rubrocuspidata		D	
Ficus septica		D	
Ficus sinuata		D	
Ficus stolonifera		D	
Ficus sumatrana		D	
Ficus sundaica	M		
Ficus uncinulata	M		
Ficus uniglandulosa		D	I
Ficus virens	M		
Parartocarpus bracteatus		D	
Prainea scandens	M		
Streblus glaber		D	

MUSACEAE

Musa beccarii		D	
Musa borneensis	M		
Musa textilis	M		I

MYRICACEAE

Myrica sp. M

MYRISTICACEAE

Gymnacranthera forbesii	M		
Horsfieldia borneensis	M		
Horsfieldia grandis	M		
Horsfieldia polyspherula	M		
Horsfieldia splendida		D	
Horsfieldia sucosa			I
Knema cinerea	M	D	
Knema conferta	M		
Knema curtisii	M		
Knema elmeri	M		
Knema galeata	M		
Knema kinabaluensis	M		
Knema korthalsii subsp. *rimosa*			I
Knema latericia subsp. *albifolia*	M	D	I
Knema latericia subsp. *ridleyi*	D		
Knema latifolia	M		I
Knema laurina		M	
Knema oblongata subsp. *oblongata*	M	D	I
Knema pallens	M		
Knema stylosa		D	
Myristica cinnamomea	M		
Myristica malaccensis	M		
Myristica maxima			I

MYRSINACEAE

Ardisia colorata	M	D	I
Ardisia crenata		D	
Ardisia cf. *elliptica*	M		
Ardisia forbesii	M		
Ardisia lanceolata	M		
Ardisia macrocalyx	M		I
Ardisia macrophylla		D	
Ardisia oxyphylla	M		
Ardisia pachysandra		D	
Ardisia platyilada		D	
Ardisia polysticta	M		
Ardisia pterocaulis	M		
Ardisia ridleyi	M		
Ardisia sanguinolenta	M		
Embelia coriacea	M		
Embelia forbesii	M		
Embelia minutifolia	M		
Embelia myrtillus	M		
Embelia oblongata	M		
Embelia philippinensis	M		
Embelia sp.		D	
Labisia pumila	M	D	I
Labisia puncata	M	D	
Maesa denticulata		D	
Maesa macrocarpa		D	
Maesa macrothyrsa	M		I
Maesa striata		D	
Maesa sumatrana	M		I

MYRTACEAE

Cleistocalyx barringtonioides		D	
Cleistocalyx perspicuinervis		D	
Decaspermum parviflorum		D	
Eugenia sandakanense		D	
Rhodamnia cinerea	M		
Rhodomyrtus sp.	M		
Syzygium alcinae	M		
Syzygium cf. *ampullaris*	M		
Syzygium bankensis	M		
Syzygium barringtoniodes	M	D	I
Syzygium calabatum	M		
Syzygium castaneum		D	
Syzygium caudatilimbum	M		
Syzygium cerasiformis	M	D	I
Syzygium chrysantha	M	D	
Syzygium claviflora	M		

	M	D	I
Syzygium cleistocalyx			I
Syzygium corymbifera	M		
Syzygium creaghii		D	
Syzygium densiflora	M		
Syzygium elliptilimba	M		
Syzygium elopurae		D	
Syzygium foxworthianum		D	
Syzygium heteroclada		D	
Syzygium hirtum	M	D	
Syzygium jambos		D	
Syzygium kalahiense		D	
Syzygium kiauense		D	
Syzygium kinabaluensis	M		
Syzygium kingii	M		I
Syzygium leucodadum		D	
Syzygium lineatum		D	
Syzygium malanccense		D	
Syzygium medium	M	D	
Syzygium myrtillus	M		
Syzygium napiformis		D	
Syzygium ochneocarpa	M		I
Syzygium paucipunctata		D	
Syzygium penibukanense		D	
Syzygium perpunticulata	M		
Syzygium rajagense	M		
Syzygium rostrata	M		
Syzygium rugosa	M		
Syzygium silamense		D	
Syzygium stapfiana	M		
Syzygium tawahense		D	
Syzygium tetragonocladum	M		I
Syzygium valdevenosa	M		
Tristania anomala	M		
Tristania grandifolia	M		
Tristaniopsis clementis	M		
Tristaniopsis obovata	M		
Tristaniopsis whiteana	M		
Tristaniopsis sp.			I
Xanthomyrtus sp.			I

NEPENTHACEAE

	M	D	I
Nepenthes gracilis	M		
Nepenthes hirsuta	M		I
Nepenthes lowii	M		
Nepenthes macrovulgaris		D	
Nepenthes cf. *mirabiles*	M		
Nepenthes rafflesiana		D	
Nepenthes reinwardtiana	M		
Nepenthes stenophylla	M		
Nepenthes tentaculata	M	D	
Nepenthes veitchii	M		

OCHNACEAE

	M	D	I
Euthemis leucocarpa	M	D	
Euthemis minor	M		
Ouratea (*Gomphia*) *serrata*	M		I
Sauvagesia (*Neckia*) *serrata*	M		I

OLACACEAE

	M	D	I
Ochanostachys amentacea	M	D	
Scorodocarpus borneensis	M	D	

OLEACEAE

	M	D	I
Chionanthus calophyllus	M		
Chionanthus crispus	M	D	
Chionanthus curvicarpus	M	D	
Chionanthus pluriflorus		D	
Chionanthus polygamus	M		
Chionanthus pubicalyx		D	
Chionanthus sp.			I
Jasminum crassifolium		D	
Jasminum melastomifolium	M		

ONAGRACEAE

	M	D	I
Ludwigia hyssopifolia	M	D	

ORCHIDACEAE

	M	D	I
Acriopsis gracilis	M		
Acriopsis liliifolia var. *auriculata*	M		
Agrostaphyllum longifolium	M		
Agrostaphyllum stipulatum	M		
Aphyllorchis montana	M	D	
Aphyllorchis pallida	M		I
Apotasia nuda	M		I
Apotasia wallichii	M	D	
Appendicula buxifolia			I
Appendicula fractiflexa	M		
Appendicula torta	M		
Appendicula undulata		D	
Arachnis calcarata			
subsp. *longisepala*		D	
Arundina graminifolia	M		
Bromheadia borneensis			I
Bromheadia finlaysoniana	M		
Bromheadia tenuis	M		
Bulbophyllum acuminatum	M		
Bulbophyllum apodum	M	D	
Bulbophyllum caudatispalum	M		
Bulbophyllum compressum		D	
Bulbophyllum farinulentum			
subsp. *farinulentum*	M		
Bulbophyllum flavescens	M		
Bulbophyllum grandilabre	M		
Bulbophyllum lobbii	M		
Bulbophyllum microglossum	M		
Bulbophyllum osyriceroides		D	
Bulbophyllum polygaliflorum	M		
Bulbophyllum pulchellum	M		
Bulbophyllum rugosum		D	
Bulbophyllum vermiculare		D	
Calanthe sollingeri		D	
Calanthe triplicata		D	
Calanthe zollingeri		D	
Callostylis leptocarpa			I
Callostylis mutans	M		
Callostylis robusta	M		
Campanulorchis discolor		D	
Campanulorchis pellipes		D	
Chelonistele lurida	M		
Claderia viridiflora			I
Coelogyne asperata	M		
Coelogyne cuprea	M		
Coelogyne pandurata	M		
Coelogyne septemcostata			I
Coelogyne aff. *rigidiformis*		D	
Cordiglottis filiformis	M		
Corybas carinatus		D	
Corybas serpentinus		D	
Cryptostylis sp.	M		
Cystorchis macrophysa		D	
Cystorchis saprophytica	M		
Cystorchis variegata		D	
Dendrobium acerosum		D	
Dendrobium aloifolium	M	D	

Species	M	D	I
Dendrobium cinnabarimum	M		
Dendrobium crumenatum	M		
Dendrobium leonis	M		
Dendrobium patentilobum	M	D	
Dendrobium puberulilingue		D	
Dendrobium sculptum	M		
Dendrochilum angustipetalum	M		
Dendrochilum cruciform	M		
Dendrochilum graminoides	M		
Dendrochilum kingii var. *kingii*		D	
Dendrochilum lambii	M		
Dendrochilum pachyphyllum	M		
Dendrochilum patentilobum	M		
Dilochia cantleyi	M		
Dimorphorchis lobbii	M		
Diplocaulobium pictum		D	
Epigeneium sp.	M		
Eulophia spectabilis	M		
Galeola nudifolia		D	I
Gastrochilus patinatus		D	
Hetaeria obliqua		D	
Lecanorchis malaccensis	M		I
Lecanorchis multiflora			I
Liparis grandiflora	M		
Liparis lacerata			I
Liparis wrayii		D	
Liparis sp.		D	
Malaxis lowii		D	
Malaxis punctatum		D	
Malleola poringensis		D	
Mischobulbum scapigerum	M		
Neuwiedia zollingeri var. *javanica*	M		
Oberonia sp.		D	
Phaius tankervilleae		D	
Phalaenopsis maculata		D	I
Phalaenopsis modesta			I
Pinalia floribunda		D	
Pinalia ignea	M		
Podochilus lucescens	M	D	
Podochilus microphyllus		D	
Porphyrodesme sarcanthoides		D	
Renanthera bella		D	
Sarcoglyphis potamophila		D	
Schoenorchis buddleiflora		D	
Selenipedium aequinoctiale			I
Spathoglottis aurea	M		
Spathoglottis kimballiana	M		
Spathoglottis microchilina		D	
Spathoglottis plicata		D	
Tainia vegetissima	M		
Thelasis pygmaea		D	
Thrixspermum centipeda	M		
Trichotosia sp.	M		
Tropidia connata			I
Vanilla sp.	M		
Vrydagzynea argentistriata	M		
Zeuxine violascens			I

OXALIDACEAE

Species	M	D	I
Dapania grandifolia		D	
Sarcotheca diversifolia	M		

PALMAE

Species	M	D	I
Areca kinabaluensis	M		
Areca minuta	M	D	I
Arenga borneensis		D	
Arenga undullatifolia	M		
Borassodendron sp.	M		
Calamus amplijugus		D	
Calamus blumei	M	D	
Calamus caesius	M		
Calamus conirostris	M		
Calamus convallium	M		
Calamus diepenhorstii	M	D	
Calamus elopurensis		D	
Calamus flabellatus	M		
Calamus cf. *gonospermus*	M		
Calamus hepburnii	M		
Calamus javensis	M	D	
Calamus laevigatus	M		
Calamus laevigatus var. *mucronatus*			I
Calamus marginatus	M	D	
Calamus mesilauensis		D	
Calamus microsphaerion		D	
Calamus muricatus	M		I
Calamus ornatus	M		I
Calamus oxleyanus	M		
Calamus pandanosmus	M		
Calamus paspalanthus	M		
Calamus pillusilus			I
Calamus pilosellus		D	
Calamus pogonacanthus	M		
Calamus praetermissus	M	D	
Calamus sarawakensis	M		
Calamus scipionum	M		
Calamus subinermis		D	
Calamus tenompokensis		D	
Calamus zonatus		D	
Caryota mitis	M		
Ceratolobus concolor	M		
Ceratolobus subangulatus			I
Daemonorops didymophylla	M		
Daemonorops elongata	M	D	
Daemonorops fissa	M	D	
Daemonorops ingens		D	I
Daemonorops korthalsii	M		
Daemonorops longipes	M	D	
Daemonorops longistipes	M		
Daemonorops microstachys	M	D	
Daemonorops oxycarpa	M		
Daemonorops ruptilis	M		
Daemonorops sabut	M	D	
Daemonorops sparsiflora	M	D	
Eugeissona utilis	M		
Iguanura cf. *polymorpha*	M		
Iguanura myochodoides		D	
Iguanura sp.			I
Korthalsia concolor	M	D	
Korthalsia echinometra	M		
Korthalsia ferox	M		
Korthalsia furtadoana	M	D	
Korthalsia jala	M		
Korthalsia rigida	M		
Korthalsia robusta	M	D	
Licuala grandis			I
Licuala longipes	M		
Licuala sabahana	M		
Licuala valida	M	D	I
Nenga gajah		D	
Oncosperma horridum	M		
Pholidocarpus maiadum	M		
Pinanga aristata	M		
Pinanga lepidota	M		
Pinanga salicifolia	M		I

	M	D	I
Pinanga variegata		D	
Plectocomia elongata	M		
Plectocomia mulleri	M	D	
Plectocomiopsis geminiflora	M		
Retispatha dumetosa	M		
Salacca cf. *affinis*	M		
Salacca clemensiana		D	
Salacca ramosiana	M		

PANDACEAE

	M	D	I
Galearia fulva	M		
Galearia sp.	M	D	I

PANDANACEAE

	M	D	I
Freycinetia ciliaris		D	
Freycinetia meijeri		D	
Freycinetia sp.	M		
Pandanus basilocularis	M		
Pandanus epiphyticus		D	
Pandanus gibbsianus	M		
Pandanus matthewsii	M	D	
Pandanus pachyphyllus		D	
Pandanus pumilus			I

PASSIFLORACEAE

	M	D	I
Adenia macrophylla var. *macrophylla*	M	D	I
Passiflora foetida		D	

PIPERACEAE

	M	D	I
Peperomia sp.	M	D	I
Piper carinum		D	
Piper pothomorphe	M		
Piper salticola		D	
Piper vestitum	M		I

PITTOSPORACEAE

	M	D	I
Pittosporum ferrugineum	M		
Pittosporum resiniferum	M		
Pittosporum silamense		D	

PODOCARPACEAE

	M	D	I
Dacrycarpus imbricatus	M		
Dacrycarpus imbricatus var. *patulus*	M		
Dacrydium beccarii	M		
Dacrydium elatum	M		
Dacrydium pectinatum	M		
Dacrydium sp.			I
Falcatifolium falciforme	M		
Phyllocladus hypophyllus	M		I
Podocarpus borneensis		D	
Podocarpus neriifolius	M		

POLYGALACEAE

	M	D	I
Epirixanthes papuana		D	
Epirixanthes sp.	M		I
Polygala paniculata		D	
Polygala sp.	M		
Xanthophyllum adenotus var. *adenotus*		D	
Xanthophyllum beccarianum	M		
Xanthophyllum discolor subsp. *discolor*	M		
Xanthophyllum flavescens	M	D	I
Xanthophyllum griffithii var. *angustifolium*	M		
Xanthophyllum havilandii	M	D	
Xanthophyllum lineare		D	
Xanthophyllum marcophyllum		D	

	M	D	I
Xanthophyllum pulchrum		D	
Xanthophyllum purpureum	M		
Xanthophyllum reticulatum	M	D	I
Xanthophyllum rufum	M		
Xanthophyllum stipitatum	M		
Xanthophyllum velutinum	M		
Xanthophyllum vittelinum		D	
Xanthophyllum sp. A (Soepadmo *et al.* 2007)		D	

PROTEACEAE

	M	D	I
Helicia attenuata	M		I
Helicia excelsa	M		
Helicia petiolaris	M		
Helicia robusta var. *robusta*	M		
Heliciopsis artocarpoides	M	D	

RAFFESIACEAE

	M	D	I
Rafflesia keithii		D	
Rafflesia tengku-adlinii	M		

RHAMNACEAE

	M	D	I
Ventilago sp.	M		
Ziziphus borneensis	M		
Zizyihus calophylla	M		
Ziziphus havilandii	M		
Ziziphus horsfieldii		D	

RHIZOPHORACEAE

	M	D	I
Anisophyllea beccariana		D	
Anisophyllea corneri	M		I
Carallia borneensis		D	
Carallia brachiata	M	D	
Gynotroches axillaris			I

ROSACEAE

	M	D	I
Angelesia cf. *splendens*	M		
Prunus arborea	M		
Prunus arborea var. *densa*	M		
Prunus arborea var. *stipulatea*	M		
Prunus polystachys	M		
Prunus spicata		D	
Prunus sp.			I
Rubus glomeratus	M		
Rubus moluccanus var. *discolor*		D	
Rubus moluccanus var. *moluccanus*	M	D	
Rubus pyrifolius		D	

RUBIACEAE

	M	D	I
Acranthera cf. *atropella*	M		
Acranthera frutescens		D	
Acranthera multiflora		D	
Acranthera velutinervia		D	
Acranthera sp.			I
Aidia borneensis	M		I
Anthocephalus chinensis	M		
Antirhea sp.	M	D	I
Argostemma boragineum	M		
Argostemma diversifolium		D	
Argostemma ophirense			I
Bungarimba sp.	M	D	I
Canthium confertum		D	
Canthium sp.	M		
Cephaelis sp.	M		
Ceriscoides imbakensis			I
Ceriscoides sp.	M	D	
Chasalia sp.	M	D	

Species	M	D	I
Coelospermum	M	D	I
Coprosma sp.	M	D	I
Coptosapelta sp.	M	D	I
Cowiea borneensis	M	D	
Crobylanthe sp.	M	D	I
Cyanoneuron pubescens	M		
Diplospora malaccensis		D	
Diplospora sp.		D	
Discospermum abnorme	M		
Discospermum sp.	M	D	I
Exallage sp.	M	D	I
Gaertnera borneensis	M		
Gaertnera vaginans	M		I
Gaertnera vaginans subsp. junghuhniana	M	D	
Gardenia tubifera	M		
Gardeniopsis sp.	M	D	I
Geocardia herbacea		D	
Geophila sp.	M	D	
Guettarda sp.	M	D	I
Gynochthodes sp.	M	D	I
Hedyotis congesta		D	
Hedyotis nitida		D	
Hedyotis cf. philippinensis	M		
Hedyotis rigida	M		
Hedyotis tenelliflora	M		
Hedyotis sp.			I
Hypobathrum sp.	M	D	I
Hypobathrum formicarium	M		
Ixora blumei	M	D	I
Ixora brachyantha		D	
Ixora cf. urophylla	M		
Ixora congesia	M		
Ixora elliptica	M		
Ixora fucosa	M		
Ixora grandiflora	M		
Ixora javanicum	M		
Ixora pyrantha	M		
Ixora stenophylla	M	D	
Ixora stenura		D	
Ixora cf. urophylla	M		
Knoxia sp.	M	D	I
Lasianthus borneensis	M	D	
Lasianthus chrysens	M		
Lasianthus inaequalis	M	D	
Lasianthus membranacens	M	D	I
Lasianthus polycarpus	M		
Lasianthus stercoriarus		D	
Lasianthus subinaequalis			I
Lecananthus sp.	M	D	I
Ludekia borneensis		D	
Maschalocorymbus corymbosus		D	
Metadina sp.	M	D	I
Mitragyna sp.	M	D	I
Morinda sp.		D	
Motleyia borneensis	M		
Mussaenda elmeri	M		
Mussaenda frondosa		D	
Mussaenda laxiflora		D	
Mussaendopsis sp.	M	D	I
Mussaluola sp.	M		
Mycetia sp.		D	
Myrmecodia sp.	M	D	I
Myrmeconauclea stipulacea		D	
Myrmeconauclea strigosa	M	D	I
Nauclea griffithii	M		
Nauclea officinalis	M		

Species	M	D	I
Nauclea subdita	M		
Neonauclea excelsioides	M		
Neonauclea gigantea		D	
Neonauclea gigantifolia	M		
Neonauclea longipedunculata	M	D	I
Neonauclea pseudocalycina	M		
Nertera sp.	M	D	I
Nostolachma sp.	M	D	I
Ochreinauclea sp.	M	D	I
Oldenlandia diffusa			I
Oldenlandia sp.	M	D	I
Ophiorrhiza winkleri	M		
Ophiorrhiza sp.		D	
Paederia verticillata	M		I
Pavetta sp.		D	I
Petunga coniocarpa			I
Phyllocrater sp.	M	D	I
Pleiocarpidia paniculata		D	
Pleiocarpidia polyneura	M		
Porterandia anisophylla	M	D	
Praravinia borneensis	M		I
Praravinia sericotricha	M		
Praravinia suberosa	M	D	
Praravinia verruculosa	M	D	
Prismatomeris beccariana	M	D	
Prismatomeris tetrandra	M	D	
Prismatomeris sp.			I
Psychotria agamae		D	
Psychotria aurantiaca	M		I
Psychotria cuspidella		D	
Psychotria densifolia	M		
Psychotria elmeri		D	
Psychotria nieuwenhuisii		D	
Psychotria sarmentosa	M		
Psychotria valetonii	M	D	
Rennellia borneensis	M		
Rennellia sp.		D	I
Richardia brasiliensis	M	D	I
Rubia sp.	M	D	I
Saprosma arborea	M		
Saprosma sp.		D	
Schradera korthalsiana	M		
Schradera montana	M		
Schradera nervulosa	M		
Scyphiphora sp.	M	D	I
Spermacoce sp.			I
Steenisia sp.	M		
Stichianthus sp.	M	D	I
Streblosa bracteata		D	
Tarennoidea sp.	M	D	I
Tarenna cumingiana	M		
Tarenna sp.		D	
Timonius eskerianus	M		
Timonius flavescens	M		
Uncaria attenuata		D	
Uncaria borneensis		D	
Uncaria calopylla	M		
Uncaria canescens		D	
Uncaria cordata	M	D	
Uncaria gambir	M		
Uncaria glabrata		D	
Uncaria jasminiflora		D	
Uncaria lanosa		D	
Urophyllum arboreum	M		
Urophyllum glabrum	M	D	
Urophyllum griffithianum	M	D	
Urophyllum griffithii	M		

Species	M	D	I
Urophyllum hirsutum	M		
Urophyllum cf. *pleiocapidia*	M		
Urophyllum woodii	M		
Urophyllum sp.			I
Wendlandia dasythyrsa	M		
Wendlandia sp.		D	I
Xanthophytum sp.	M	D	I
Zeuxantha moultonii	M		

RUTACEAE

Species	M	D	I
Citrus macrocarpa var. *macroptera*		D	
Clausena excavate var. *excvata*	M	D	
Glycosmis chlorosperma var. *elmeri*		D	
Glycosmis macrantha		D	
Glycosmis sapindioides var. *sapindioides*			I
Luvunga heterophylla		D	I
Luvunga sarmentosa	M	D	I
Maclurodendron porteri	M		
Melicope confusa		D	
Melicope incana		D	
Melicope lunu-ankenda		D	
Melicope subunifoliolata	M	D	
Micromelum minutum var. *minutum*		D	
Pleiospermium latiolatum			I
Tetractomia tetandrum	M	D	

SABIACEAE

Species	M	D	I
Meliosma sumatrana	M		I
Polyosma maliauensis	M		

SANTALACEAE

Species	M	D	I
Dendrotrophe reinwardtiana	M		
Dendrotrophe varians	M	D	
Scleropyrum pentandrum	M	D	

SAPINDACEAE

Species	M	D	I
Allophyllus cobbe	M	D	I
Dimocarpus dentatus		D	
Dimocarpus longan			I
Dimocarpus longan subsp. *malesianus*		D	
Guioa pleuropteris	M		
Guioa pterorhachis	M	D	
Harpullia arborea	M	D	
Lepisanthes alata		D	
Lepisanthes sp.	M		
Mischocarpus pentapetalus	M		
Mischocarpus sundaicus	M		
Nephelium cuspidatum var. *robustum*	M		
Nephelium maingayi	M		
Nephelium ramboutan-ake	M	D	I
Nephelium uncinatum	M		
Paranephelium xestophyllum	M	D	I
Pometia pinnata	M	D	I
Xerospermum laevigatum	M		
Xerospermum noronhianum	M		

SAPOTACEAE

Species	M	D	I
Chrysophyllum roxburghii			I
Madhuca burckiana			I
Madhuca cheongiana	M		
Madhuca kingiana	M		I
Madhuca korthalsii	M	D	
Madhuca malaccensis	M		
Madhuca mindanaensis	M	D	I
Madhuca multinervia			I
Madhuca pallida			I

Species	M	D	I
Madhuca pubicalyx		D	
Madhuca sandakenensis	M		
Madhuca silamensis		D	
Palaquium beccarianum	M		
Palaquium calophyllum			I
Palaquium dasyphyllum	M		I
Palaquium gutta	M		I
Palaquium leiocarpum	M		
Palaquium rostratum	M		
Palaquium sericeum	M	D	
Payena acuminata		D	
Payena gigas	M		
Payena microphylla	M		

SAXIFRAGACEAE

Species	M	D	I
Polyosma cyanea	M		
Polyosma integrifolia	M		
Polyosma latifolia	M		
Polyosma mutabilis	M		

SCROPHULARIACEAE

Species	M	D	I
Brookea dasyanthea		D	
Brookea tomentosa		D	
Brookea sp.	M		
Cyrtandromoea grandes			I
Lindernia crustacea		D	I
Lindernia ruelloides	M	D	
Lindernia viscosa			I
Torenia peduncularis	M	D	
Torenia violacea		D	
Hygrophila sp.	M		

SCYPHOSTEGIACEAE

Species	M	D	I
Scyphostegia borneensis	M	D	I

SIMAROUBACEAE

Species	M	D	I
Eurycoma longifolia	M	D	I

SMILACACEAE

Species	M	D	I
Smilax borneensis	M	D	
Smilax laevis	M		

SOLANACEAE

Species	M	D	I
Lycianthes biflora var. *mollisima*	M		
Lycianthes biflora		D	
Lycianthes nigrum		D	
Lycianthes parasitica	M	D	I
Solanum ferox		D	I

SONNERATIACEAE

Species	M	D	I
Duabanga moluccana	M	D	I

SPARMANIACEAE

Species	M	D	I
Brownlowia peltata	M	D	I
Grewia acuminata		D	
Grewia laevigata		D	
Microcos antidesmifolia	M		
Microcos antidesmifolia var. *hirsuta*	M		
Microcos cinamomifolia	M		
Microcos elmeri	M		
Microcos latistipulata var. *latistipulata*		D	
Microcos opaca		D	
Microcos ossea		D	
Microcos pearsonii		D	
Microcos reticulata	M	D	
Microcos triflora var. *triflora*		D	
Mocrocos latistipulata var. *latistipulata*		D	

Species	M	D	I
Pentace adenophora			I
Pentace borneensis			I
Pentace laxiflora	M	D	

STERCULIACEAE

Species	M	D	I
Byttneria sp.		D	
Commersonia bartramia		D	
Firmiana malayana		D	
Heritiera borneensis	M		
Heritiera elata	M		
Heritiera impressinervia	M		
Heritiera simplicifolia	M		
Heritiera sumatrana	M		
Leptonychia caudata	M		
Pterospermum javanicum		D	
Pterospermum oblongum	M		
Pterospermum stapfianum		D	
Scaphium affine	M		
Scaphium longipetiolatum	M		
Scaphium macropodium	M		
Sterculia coccinea		D	
Sterculia cordata	M		
Sterculia rubiginosa		D	
Sterculia rubiginosa var. *setistipula*	M		
Sterculia stipulata	M	D	I

STYRACACEAE

Species	M	D	I
Bruinsmia styracoides	M	D	

SYMPLOCACEAE

Species	M	D	I
Symplocos anomala	M		
Symplocos brachybotrus	M		
Symplocos buxifolia	M		
Symplocos confusa		D	
Symplocos fasciculata		D	
Symplocos henschelii	M		
Symplocos odoratissima var. *wenzelii*		D	
Symplocos ophirensis	M		
Symplocos pendula var. *hirtistylis*	M		
Symplocos tricoccata		D	

THEACEAE

Species	M	D	I
Adinandra acuminata		D	
Adinandra clemensiae	M	D	
Adinandra collina	M		
Adinandra cordifolia	M		I
Adinandra dumosa	M		
Adinandra excelsa	M		
Adinandra miquelianus	M		
Adinandra subsessilis			I
Camelia sp.			I
Eurya sp.		D	
Eurya acuminata	M		
Eurya obora	M		
Gordonia sarawakensis	M		
Pyrenaria kunstleri	M		
Pyrenaria parviflora	M		
Pyrenaria tawauensis	M	D	I
Schima breviflora	M		
Schima monticola	M		
Schima wallichiana	M		
Schima wallichii subsp. *wallichii*	M	D	
Schima wallichii subsp. *monticola*	M		
Ternstroemia aneura	M		
Ternstroemia coriacea `	M		
Ternstroemia elongata	M		
Ternstroemia cf. *micocalyx*	M		

Species	M	D	I
Ternstroemia sp.			I
Tetramerista glabra	M		

THYMELAEACEAE

Species	M	D	I
Aquilaria beccariana		D	
Aquilaria malaccensis	M	D	
Gonystylus bancanus	M		
Gonystylus forbesi	M		
Gonystylus keithii		D	
Phaleria capitata		D	
Phaleria octandra		D	
Wikstroemia androsaemifolia	M		
Wikstroemia brachyantha	M		
Wikstroemia polyantha	M		
Wikstroemia tenuiramis	M		

TRIGONIACEAE

Species	M	D	I
Trigoniastrum hypoleucum	M	D	

ULMACEAE

Species	M	D	I
Gironniera nervosa	M		
Gironniera subaequalis	M		

URTICACEAE

Species	M	D	I
Astrothalamus sp.	M		
Dendrocnide oblanceolata		D	
Dendrocnide stimulans		D	
Elatostema acuminata		D	
Elatostema integrifolium	M		I
Elatostema variolaminosum		D	
Leucosyke capitellata var. *eucapitellata*		D	
Leucosyke capitellata			I
Pilea sp.		D	I
Pipturus argenteus		D	
Pipturus cf. *asper*		D	
Poikilospermum oblongifolium		D	
Poikilospermum cordifolium		D	
Poikilospermum microstachys		D	I
Poikilospermum scortechinii	M	D	
Poikilospermum suaveolens	M	D	
Poikilospermum tangaum			I

VERBENACEAE

Species	M	D	I
Lantana camara	M	D	I
Stachytarpheta sp.	M	D	I

VIOLACEAE

Species	M	D	I
Rinorea anguifera	M		
Rinorea bengalensis		D	
Rinorea congesta			I
Rinorea iliaspaiei		D	
Rinorea longinacemosa			I

VISCACEAE

Species	M	D	I
Viscum ovalifolium		D	
Viscum wrayi	M		

VITACEAE

Species	M	D	I
Ampelocissus imperialis	M		
Ampelocissus ocheacca		D	I
Cayratia sp.		D	
Cissus adnata		D	
Cissus angustata	M	D	I
Cissus angustifolia		D	
Cissus discolor		D	
Cissus hastata	M	D	I
Cissus nodosa		D	

	M	D	I
Cissus repens			I
Cissus simplex	M	D	I
Pterisanthes cissoides		D	
Pterisanthes sp.	M		
Pterisanthes quinquefolialata			I
Tetrastigma dichotomun	M		
Tetrastigma diepenhostii	M	D	I
Tetrastigma dubium	M		
Tetrastigma lanceolarium	M		
Tetrastigma magnum		D	
Tetrastigma papillosum	M		
Tetrastigma pedunculare	M	D	I

WINTERACEAE

	M	D	I
Drimys piperata	M		

ZINGIBERACEAE

	M	D	I
Achasma sp.	M		
Alpinia assimile		D	
Alpinia fraseriana	M		
Alpinia glabra			I
Alpinia havilandii			I
Alpinia ligulata	M	D	I
Amomum anomalum			I
Amomum borealiborneense			I
Amomum dimorphum			I
Amomum hansenii		D	
Amomum laxisquamosum		D	I
Amomum oliganthum		D	I
Amomum polycarpum	M		
Amomum staminidivum			I
Amomum uliginosum			I
Boesenbergia aurantiaca		D	
Boesenbergia pulchella		D	I
Boesenbergia sp.	M		
Burbidgea pubescens	M	D	I
Burbidgea schizocheila		D	
Cenolophon sp.	M		
Cotylanthera tenuis	M		
Elettaria longituba			I
Elettaria sp.	M		
Elettariopsis sp.	M		I

	M	D	I
Etlingera albolutea		D	
Etlingera aurantia	M	D	I
Etlingera baculutea	M		
Etlingera belalongensis		D	
Etlingera brachychila var. *vinosa*			I
Etlingera brevilabrum	M		I
Etlingera coccinea	M	D	I
Etlingera corrugata		D	
Etlingera dictyota	M		
Etlingera fimbriobracteata	M	D	
Etlingera inundata		D	I
Etlingera megalocheilos		D	
Etlingera pubescens		D	I
Etlingera rosamariae	M		
Etlingera rubromarginata	M	D	I
Etlingera sessilanthera	M	D	I
Geocharis sp.		D	
Geostachys maliauensis	M		
Globba atrosanguinea	M	D	I
Globba fraciscii	M	D	
Globba macrocarpa		D	
Globba pendula	M	D	I
Globba propinqua	M		
Globba tricolor		D	
Goecharis fusiformis var. *borneensis*			I
Hedychium cf. *cylindricum*	M		
Hedychium muluense		D	
Hedychium sp.			I
Hornstedtia havilandii			I
Hornstedtia scyphifera		D	
Hornstedtia sp.	M		
Languas galanga	M		
Plagiostachys albiflora	M	D	I
Plagiostachys breviramosa		D	I
Plagiostachys parva	M	D	I
Plagiostachys strobilifera	M	D	I
Zingiber alboflorum			I
Zingiber coloratum	M		
Zingiber flagelliforme		D	
Zingiber lambii		D	
Zingiber pachysiphon			I
Zingiber vinosu	M		

Acknowledgements

We would like to thank our many collaborators: in Sabah at the Herbarium of the Forest Department, John B. Sugau, Joan T. Pereira, Suzana Sabran, Postar Jaiwit Miun and Rosalia Eson; in Kota Kinabalu, Yayasan Sabah Group, Mohd Daud Tampokong and Waidi Sinun; in Danum Valley, Glen Reynolds, Alex Karolus, Bernadus Bala Ola, Ismail Bin Samat, Philip Ulok, Jamil Kasmin and F. T. Welly; and at the Royal Botanic Gardens, Kew, Suzanna Baena (fieldwork), Bill Baker (palms and Pandanaceae), Henk Beentje (Compositae), Renata Borosova (scanning), Gemma Bramley (Gesneriaceae, Lamiaceae, Verbenaceae and fieldwork), Marie Briggs (fieldwork), Chill Chalan (Euphorbiaceae), Ruth Clark (legumes), Aaron Davis (Rubiaceae), Clare Drinkell (scanning), Jonathan Gregson (fieldwork), Nicolas Hind (Compositae), Don Kirkup (Loranthaceae), Gwil Lewis (legumes), Eve Lucas (Melastomataceae and Myrtaceae), Sophie Marsh (fieldwork); Alison Moore (Moraceae and fieldwork), Melanie Thomas (Moraceae, Ulmaceae and Urticaceae), Jovita Yesilyurt (fieldwork), Kaj Vollesen (Acanthaceae), Soraya Villalba (scanning), Maria Vorontsova (Solanaceae), Paul Wilkin (various monocot families), Jeff Wood (Orchidaceae) and Sue Zmarzty (Flacourtiaceae).

We would also like to thank The UK Government Darwin Initiative for funding this project and the Tobu Foundation for supporting some of the fieldwork.

© A. McRobb/RBG Kew

Index

The most commonly encountered families and the page numbers of their descriptions are shown in **bold**.